Underwater Tailing Placement
at
Island Copper Mine

A SUCCESS STORY

George W. Poling, Derek V. Ellis, James W. Murray,
Timothy R. Parsons, and Clem A. Pelletier

Published by the
Society for Mining, Metallurgy, and Exploration, Inc.

*In addition to the five authors, many others collected and evaluated the
data described in this book. We are particularly grateful to the
management and staff of the Island Copper Mine, who helped us
tremendously over the last 31 years. Without their assistance and
cooperation, we would not have been able to compile and publish this case
study. Similar thanks go to the many graduate students and colleagues at
the University of British Columbia, the University of Victoria, Simon
Fraser University, and Rescan Environmental Services, Ltd., who
contributed a great deal to this work. And to BHP Billiton, we give sincere
thanks for supporting all these efforts for the last three decades.*

Society for Mining, Metallurgy, and Exploration, Inc. (SME)
8307 Shaffer Parkway
Littleton, Colorado, USA 80127
(303) 973-9550 / (800) 763-3132
www.smenet.org

SME advances the worldwide mining and minerals community through information
exchange and professional development.

ISBN 0-87335-214-9

Library of Congress Cataloging-in-Publication Data

Underwater tailing placement at Island Copper Mine / George W. Poling ... [et al.].
    p. cm.
Includes bibliographical references and index.
ISBN 0-87335-214-9
   1. Mineral industries--Waste disposal--Environmental aspects. 2. Waste disposal in the
ocean. 3. Tailings (Metallurgy)--Environmental aspects. I. Poling, George W.

TD899.M47 U53 2002
622'.028'6--dc21

                                                      2002070632

Contents

LIST OF FIGURES vii

LIST OF TABLES xi

CHAPTER 1   **AN INTRODUCTION TO DEEP SEA TAILING PLACEMENT** 1
George W. Poling

Brief History of Deep Sea Tailing Placement 2

Brief Overview of Deep Sea Tailing Placement Systems 3

Brief Description of the Deep Sea Tailing Placement Technology 5

Overview of Biological Concerns 7

Comparison Between Deep Sea Tailing Placement
and On-Land Impoundment 9

Advantages of Deep Sea Tailing Placement Systems 9

Disadvantages of Deep Sea Tailing Placement Systems 11

Conclusions 12

References 14

CHAPTER 2   **SELECTION OF SUBSEA TAILING PLACEMENT** 17
George W. Poling

Exploration and Discovery 17

Permitting the Mine 19

Preoperational Baseline Data 19

Public Hearing 24

Monitoring of Environmental Impacts 27

Problems and Controversies During Mine Life 30

References 32

CHAPTER 3   **ENGINEERING CHALLENGES AND SOLUTIONS** 35
George W. Poling

Mining 36

Mineral Processing 38

Capital and Operating Costs 43

References 43

CHAPTER 4   **THE HISTORY OF THE MORPHOLOGICAL CHANGE ON THE SEAFLOOR OF RUPERT AND HOLBERG INLETS 45**
James W. Murray and Clem A. Pelletier

Background **45**

Purpose **47**

Field Instrumentation and Surveys **49**

Data Processing and Map Preparation **51**

Results and Discussion **58**

Summary and Conclusions **81**

Acknowledgments **82**

References **82**

CHAPTER 5   **GEOCHEMISTRY: CHEMICAL STABILITIES OF TAILINGS SEDIMENT 85**
George W. Poling

Mineralogical Composition of Island Copper Mine Tailings Solids **85**

Tailings as Marine Sediments **86**

Bioassays on Tailings **90**

Seawater Column Chemical Quality **91**

References **91**

CHAPTER 6   **CHANGES IN PHYSICAL AND CHEMICAL PROPERTIES OF RUPERT INLET WATERS 93**
Timothy R. Parsons

Introduction **93**

Methods **94**

Results and Discussion **95**

Comparison of Data During Mine Operations **95**

Summary **102**

Acknowledgments **103**

References **103**

CHAPTER 7   **CHANGES IN THE BIOLOGICAL PROPERTIES OF THE PELAGIC ENVIRONMENT OF RUPERT INLET WATERS DURING 22 YEARS OF MINE OPERATIONS 105**
Timothy R. Parsons

Introduction **105**

Methods **106**

Results and Discussion **106**

Summary **113**

Acknowledgments **113**

References **113**

CHAPTER 8   **SEABED BIODIVERSITY AT ISLAND COPPER MINE: IMPACT AND RECOVERY 115**
Derek V. Ellis

Introduction **115**

The Sediment Benthos (Infauna) Surveys **116**

The Sediment Habitat and Biological Interactions **120**

The Diversity of Species (Species Richness) **123**

The Numbers of Each Species (Species Evenness) **128**

The Biodiversity Expressed in Terms of High-Level Taxonomic Units **130**

Similarity and Diversity Indices **131**

Large Infaunal Species **131**

Sustainability of the Benthic Recovery from Tailings Impact **132**

Acknowledgments **133**

References **133**

CHAPTER 9   **UNDERWATER BIODIVERSITY SURVEYS AND BIOLOGICAL COLONIZATION OF THE WASTE DUMP SHORELINE 137**
Derek V. Ellis

Introduction **137**

The Underwater Surveys **139**

Initial Diver Surveys by Mine Environmental Staff and Consultants, 1972–1974 **139**

Diver Surveys by Federal Government Scientists, 1970–1975 **140**

Submersible Surveys, 1975–1984 **141**

Diver Surveys, 1983 and 1999 **142**

Conclusions Drawn from the Underwater Surveys **142**

Shoreline Colonization of the Beach Waste Dump **143**

Acknowledgments **145**

References **145**

CHAPTER 10   **FISHERIES, TAILINGS BIOASSAYS, TRACE METAL BIOACCUMULATION IN BENTHOS, AND SETTLING PLATES 147**
Derek V. Ellis

Introduction **147**

The Crab Fishery and Yield **148**

Salmon and the Salmon Hatchery **151**

Tailings Bioassays **152**

Tissue Monitoring of Trace Metal Bioaccumulation by Benthos **153**

Settlement Plates and Suspended Solids **157**

Acknowledgments **159**

References **159**

CHAPTER 11 **POSTCLOSURE REHABILITATION AND ASSESSMENT OF INLET SYSTEM 161**
George W. Poling
Post-closure Monitoring of the Marine Inlets **164**

APPENDIX A 167

APPENDIX B 177

INDEX 191

COLOR PLATES 197

Figures

1.1   Schematic of a gravity-based typical DSTP system **6**

2.1   Location of ICM along the north shore of Rupert Inlet **18**

2.2   Projected area of tailing deposition in Rupert Inlet from 25 years of discharge **22**

2.3   Schematic diagram of seawater mix tank and outfall system **31**

2.4   Photograph of the seawater mix tank at ICM (operating mix tank is nearest the viewer, with the original kept as a spare) **32**

3.1   Map of Rupert–Holberg–Quatsino inlet areas showing location of mine and marine sampling stations **36**

3.2   Geological plan view of open pit showing major rock types **37**

3.3   Geological section through open pit showing major rock types **37**

3.4   Sinking cut blast in open pit mine at Island Copper **38**

3.5   Simplified schematic flowsheet of Island Copper process plant **39**

3.6   Particle size distribution of Island Copper ground ore and tailings **40**

4.1   Rupert Inlet and ICM site **46**

4.2   Pre-mine composite bathymetry of Rupert Inlet **48**

4.3   Bathymetry of Rupert Inlet (January 2000) **50**

4.4   January 2000 sounding tracks of Rupert Inlet **52**

4.5   Locations of cross sections **53**

4.6   Sample sounding profile along cross section 6 **54**

4.7   Seismic lines across Rupert/Holberg Inlet **55**

4.8   Seismic profile through line 19 **56**

4.9   Rupert Inlet seismic profile number 19 **57**

4.10   Schematic depositional model of the mine-related materials in Rupert/Holberg Inlet **62**

4.11   Superimposed isobaths of pre-mine and January 2000 surveys **63**

4.12   Superimposed profiles of cross sections 1 and 2 **64**

4.13   Superimposed profiles of cross sections 3 and 4 **64**

4.14   Superimposed profiles of cross sections 5 and 6 **65**

4.15   Superimposed profiles of cross sections 7 and 8 **65**

4.16   Superimposed profiles of cross section 9 **66**

4.17   Isopach map of deposition of Rupert Inlet from bathymetric survey **67**

4.18   Isopach map of mine-derived material in Rupert Inlet from January 2000 seismic survey **68**

4.19   Seismic profile through line X7 across the scour hole **69**

4.20   Thickness of deposits in the confluence scour hole at Hankin Point (1976–2000) **70**

4.21   Pre-mine and present hypsometry of Rupert Inlet **71**

4.22   Pre-mine and present hypsometry of Rupert Inlet at cross sections 1 and 2 **72**

4.23   Pre-mine and present hypsometry of Rupert Inlet at cross sections 3 and 4 **73**

4.24   Pre-mine and present hypsometry of Rupert Inlet at cross sections 5 and 6 **74**

4.25   Pre-mine and present hypsometry of Rupert Inlet at cross sections 7 and 8 **75**

4.26   Pre-mine and present hypsometry of Rupert Inlet at cross section 9 **76**

4.27   Sensitivity of rock fill and tailings tonnage to change in dry density **78**

5.1   Effect of seawater addition on zeta potential **87**

5.2   Schematic distribution of biogeochemically important species in interstitial waters in sediments (Pedersen 1985) **89**

5.3   Rupert Inlet and adjacent fjords with selected monitoring stations **91**

5.4   Dissolved copper in seawater (stations A, E, and F; 1971–1993) **92**

5.5   Dissolved manganese in seawater (stations A, E, and F; 1971–1993) **92**

6.1   Station locations **94**

6.2   Average turbidity and Secchi disc depth at station A, 1971–1992, during each month of the year **96**

6.3   The bar plot with linear fit for dissolved copper, dissolved manganese, dissolved zinc, and total arsenic observed at station A from 1971–1992 **101**

7.1   Station locations **107**

7.2   The concentration of chlorophyll A observed at station A from 1971 to 1992— a bar plot with linear fit (upper left), a line plot with polynomial fit (upper right), a box-whisker plot for annual means (lower left), and a box-whisker plot for seasonal means (lower right) **109**

8.1   Example infaunal benthos: small crab, shrimp, fish, worms, and snails can be seen **117**

8.2   Station sampling positions (numbered) are located around the area targeted for tailings deposition (<100 m depth) **118**

8.3   Benthic sample equipment of the type used at ICM **119**

9.1   Scuba dive transects were located at the settlement plate sites (see Chapter 10), including the mill dock (#10) and the tailings outfall (#11) **138**

10.1   Map showing crab sampling stations **149**

10.2   Trace metals in mussels (*Mytilus edulis*) **155**

10.3   Trace metals in rockweed (genus *Fucus*) **156**

10.4   Trace metals in Dungeness crab (*Cancer magister*) **157**

10.5   Map showing sampling stations for settling plates **158**

11.1   Aerial photograph of the ICM site, taken from the west, near the time of mine closure **162**

11.2   The seawater fall viewed from the pit side during the flooding process **162**

11.3   State of land and water reclamation at ICM, June 2001 **163**

COLOR PLATES

9.1   Top: Two delicate, shell-less mollusks (nudibranchs) crawl together by an elongated tube-dwelling sea anemone on the rock at Hankin Point. The background consists of rocks with abundant algae and a patch of sediment. Bottom: A similar area at Hankin Point shows a film of sediment but with the same type of mollusk, a sea anemone, and the algal forest growing on the rocks. Both photographs were taken in July 1999. **197**

9.2   Top: This surface roughness feature has accumulated sediment where a massive slime-worm is growing. Dense algae grow on the rocks that remain exposed. Bottom: On this patch of sediment between algae-supporting rocks is one of the many fish recorded, along with a starfish that is apparently feeding on a prey organism. Both photographs were taken in July 1999 at Hankin Point. **198**

9.3   Top: This patch of sediment at Hankin Point is supporting two large burrowing sea anemones and a starfish. Bottom: A Dungeness crab is burrowed into the patch of sediment, with algae growing on the nearby exposed rock. Both photographs were taken in July 1999. **199**

9.4   Top: A supportive cable at the mine dock is crowded with growth, including a patch of tunicates, a great deal of algae, and encrusting organisms. Bottom: In the near-shore mud around the dock, substantial numbers of burrowing sea anemones are living. The veneer of brown organic ooze common on muds throughout the fjord system is visible. Both photographs were taken in July 1999. **200**

9.5   Top: Thick and diverse encrusting organisms, including one large tubeworm that is well-extended and feeding, grow on the tailings outfall. Bottom: A dense colony of sea anemones is growing on the outfall. Both photographs were taken in July 1999. **201**

9.6   Top: A fish grazes on the dense encrusting growth on the tailings outfall. Bottom: On the tailings bed by the outfall site, a starfish moves past the imprint of a smaller starfish in the brown organic ooze widely spread throughout the fjord system. Both photographs were taken in July 1999. **202**

9.7   Top: The view from the submersible port over natural (brown) sediments showing the 1-m^2 quadrat and the measuring rope. A crustacean is present in the foreground by the rope, with several burrow holes nearby. Bottom: A similar view over the tailings (gray). A small fish is present at left center, and several burrow holes are visible. **203**

9.8   Top: A large boulder near the mine site shows the typical regional shoreline zonation pattern, with its expression on the surrounding stony beach. Bottom: The shoreline front between normal rock shore zonation for a stony/boulder beach (left) and the first-year colonization of low-tide green and brown filamentous algae (right). **204**

Tables

1.1 International DSTP sites currently operating, currently proposed, recently closed, and proposed but not implemented **4**

1.2 Comparison between submarine tailing placement (STP) and on-land placement of sites with high precipitation rates **10**

2.1 Chemical analyses of various size fractions of solid portion of effluent from flotation test of ICM ore **20**

2.2 Projected 1:1 dilution effluent concentrations in Utah Construction and Mining Ltd. (1970) compared to eventual permit awarded **21**

2.3 Outline of environmental control monitoring program, ICM third production year, October 1973 to October 1974 **28**

3.1 Typical chemical and mineralogical compositions of tailing solids and natural sediments in Rupert Inlet, British Columbia **41**

3.2 Typical mill process reagents used at the ICM concentrator **42**

4.1 Bathymetric and continuous seismic profiling surveys of Rupert Inlet from 1969 to 2000 **47**

4.2 Summary of survey parameters **51**

4.3 Changes in volume/cross-sectional area of Rupert/Holberg Inlet as a result of mine-related material disposal **77**

4.4 Summary of mine-related material budget **80**

6.1 Results from the linear regression of variables over time for data from 1971 to 1992 **97**

6.2 A summary of differences between means of Secchi disc depth and turbidity before and after closure **99**

6.3 A summary of mean metal concentrations (μg/L) during the period from 1971 to 1992 **100**

6.4 Comparison of the mean of metal levels in Rupert Inlet, Holberg Inlet, and Quatsino Sound during 1993–1995 (before closure) and during 1996–1998 (after closure) **102**

6.5 Comparison of the means of metal levels at stations close to the rock dump in Rupert Inlet, 1993–1995 (before closure) and 1996–1998 (after closure) **102**

7.1 Comparison of variable means before and after September 1971 by ANOVA **108**

7.2   Correlation between variables from data obtained over the years from 1971 to 1992 **108**

7.3   Descriptive statistics for the total counts of zooplankton from vertical hauls between 1971 and 1993 **110**

7.4   Results from t-tests for the total counts of zooplankton from vertical hauls between 1971 and 1993 **111**

7.5   Correlation of total zooplankton with chlorophyll A, temperature, dissolved copper, dissolved manganese, dissolved zinc, and total arsenic **111**

7.6   Results from linear regressions for the total counts of zooplankton from vertical hauls between 1971 and 1993 **112**

7.7   Results from linear regressions for the diversity index of zooplankton **112**

7.8   Changes in chlorophyll A and zooplankton numbers before and after mining in Rupert Inlet **113**

8.1   Sampling station sediment designations based on grain size (Sheppard or Wentworth terminology as used by original authors) **121**

8.2   Number of species at each sampling station each year per unit area sampled (0.15 m^2) **125**

10.1   Summary of catches of Dungeness crab (*Cancer magister*), 1970–1998 **150**

10.2   Tailings bioassay results; acute toxicity tests (LC$_{50}$ 96-h) on juvenile salmonids **153**

10.3   Settling plate data: settling rates (g/m^2/day) for total and volatile solids **159**

CHAPTER 1

An Introduction to Deep Sea Tailing Placement

George W. Poling

In today's world, the mining industry's use of space must compete with all other land uses, such as wilderness, recreation, wetlands, ranching, agriculture, population centers, and forestry. If we focus primarily on the base metal, precious metal, and ferrous metal mining industries, sites for placement of waste rock and tailings usually occupy the largest land areas within a mining site.

Today's regulations normally require demolition of buildings and reclamation of these sites plus roads after the mine has closed. Waste rock must normally be drilled, blasted, and excavated to gain access to "ore" grade rock that is similarly mined and subsequently processed to recover the valuable minerals or metals contained therein. During the operating life of the mine, this waste rock is normally stored in piles adjacent to the mining open pit or a shaft that accessed an underground orebody. After closure, these piles are recontoured and revegetated for reclamation purposes. It is common to mine several times more waste rock than ore, particularly in an open pit mine or strip mine, so as to safely expose the ore itself. In large-scale mines, several hundred million tonnes of waste rock must be stored on many hectares of land at a mining site.

Tailings often comprise the majority of the solid and liquid products of a mineral processing operation after the valuable constituents from the ore have been physically or chemically removed. Often, tailings are finely ground solids that, together with their aqueous components, make up a "slurry." This tailing slurry can amount to hundreds of millions of tonnes during the life of a large-scale mine. To satisfy mine builders, regulators, and the public, the slurry must be safely stored or "placed," normally in a land-based impoundment. These waste rock storage piles and tailing storage facilities must remain secure (stable) and noncontaminating for all time.

When tailings are stored on land, dams or dikes are normally employed to impound slurry. Several of these rank among the largest man-made structures in the world. Communities located near these tailing impoundments understandably regard such structures as both environmental and sometimes personal risks. Perceived risks include (1) structural failure of dams or dikes with consequent downstream flooding and inundation by tailing solids, which can fluidize and sometimes flow for several kilometers downstream (Das and Poling 2000); (2) "dusting," which is air pollution caused by unconsolidated clays and sand components of the tailings during high winds; and (3) contamination of downstream surface and groundwaters by seeps or runoff from the impoundment. On-land impoundments can also be expected to be visually unattractive. In several instances, after careful analyses of tailing placement alternatives, mine proponents and government regulators have identified deep sea tailing placement (DSTP) as a technology preferable to on-land impoundment. BHP Billiton's Island Copper Mine (ICM) near the northern end of Vancouver Island on British Columbia's West Coast, which is the prime focus of this volume, is a prime example of such a regulatory decision.

Today, prospects of deep sea placement of mill tailings can apply only to a few specific mines located reasonably close to a marine coast that has deep waters close to the shore. To justify evaluation of a DSTP alternative, a mine must be within approximately 200 km of a coast. In addition, the tailings themselves must be demonstrated to not be acutely toxic at their point of discharge into the marine environment. Finally, the proposed site for tailing solid deposition, as a sediment footprint, must be of low biological resource value, and the transport route to the site and the site itself should present a very low risk of tailing solids spilling into or upwelling into the uppermost photosynthetic zone (the euphotic zone). Real and perceived risks of DSTP include alteration of the topography/bathymetry of the sea floor, inundation of bottom-dwelling creatures (benthos), possible heavy metal contamination of sea creatures, and potential adverse impacts on local fisheries. Some of the most critical ocean and shoreline resources, such as coral reefs and mangrove forests, are particularly sensitive to damage resulting from sedimentation and must be protected in all DSTP designs.

BRIEF HISTORY OF DEEP SEA TAILING PLACEMENT

Shallow placement of mill tailings has been employed around the world for more than a century with various adverse environmental impacts (Ellis, Poling, and Pelletier 1995; Ellis 1989). Some recent large-scale mines that used either riverine transport to the sea or shallow submarine tailing discharge are the Toquepala and Cuajone Mines in Peru; the Marcopper Mine in the Philippines; the Bougainville Copper Mine in Papua, New Guinea; and a small underground mine, the Jordan River Mine, on the southwest coast of Vancouver Island, British Columbia. Case histories of these mines can be found in Ellis, Poling, and Pelletier (1995).

Within the past 30 years, DSTP has emerged as an alternative tailing storage technology. This has been based on advancements in environmental monitoring and assessment methods, particularly with respect to oceanographic monitoring and modeling, as well as advanced engineering systems for pipeline transport and subsea placement of deaerated tailings at ocean depths currently up to 385 m. Beyond the pipe terminus, some DSTP systems have tailing slurries flowing as density currents to depths of several thousand meters before deposition. The ICM is one of the best-documented DSTP case studies. More than 400 million tonnes of finely ground tailing solids were placed on the sea floors of Rupert Inlet and Holberg Inlet off Vancouver Island. The ICM is the first documented case of engineered DSTP that underwent a full-blown public inquiry and detailed regulatory scrutiny before being granted a permit by the province of British Columbia in 1971. This open pit mine began production of copper sulfide and molybdenite mineral flotation concentrates in October of 1971 and operated continuously until December of 1995. At the time of writing, nearly 5 years of postclosure environmental monitoring of these two inlets and beyond have been recorded and reported to both provincial and federal governmental regulators. This case study revealed none of the perceived risks of DSTP outlined previously. Heavy metal bioaccumulation did not occur to any significant extent, and fisheries were not adversely affected. In both inlets, a commercial crab fishery and a recreational salmon fishery continued unabated throughout the life of the mine and the post-closure period. Even commercially farmed salmon finishing pens have operated along the shoreline of Holberg Inlet with no apparent adverse effects from the tailings flowing and sedimenting beneath them. Benthos biodiversity was reduced in the deepest portions of these two inlets, where rates of deposition of tailing solid particulates were high (greater than 20 cm/yr). The impact on this small proportion of the inlet benthos did not affect the overall ecology of the inlets during this mine's 25-year life. Within 1 to 3 years of mine closure, the biodiversity reductions that did occur had recovered to a sustaining ecological succession. Even heavily sedimented regions on the inlet floors were being colonized by breeding populations of primary opportunistic species within 9 months of mine closure. Much more detail on all aspects of the ICM DSTP system and marine receiving environment is provided in the remaining chapters of this book. This introductory chapter presents a brief stand-alone overview of DSTP history, technology, and permitting, along with comparisons to other tailing placement alternatives. Many of these same aspects will be covered in greater detail for the ICM case study in the following chapters.

DSTP is currently being used at six mining operations around the world. Table 1.1 itemizes these locations and several characteristics of each operation. The table also lists several other operations where DSTP is being proposed, four operations that recently closed, and four mine sites where DSTP was evaluated and not implemented for a variety of reasons. As the table shows, DSTP is a significant alternative tailing storage technology that can and is competing successfully with on-land storage alternatives in a wide variety of political jurisdictions and environmental sensitivities.

BRIEF OVERVIEW OF DEEP SEA TAILING PLACEMENT SYSTEMS

In 1992, Baer, Sherman, and Plumb published an overview and bibliography titled *Submarine Disposal of Mill Tailings from On-Land Sources—An Overview and Bibliography* as background for the Alaska Field Operation Center of the U.S. Bureau of Mines (USBM). As indicated in Table 1.1, at least three DSTP systems have been proposed and evaluated for southern Alaska's deep fjord-type coastal waters. Although none of these systems has yet become operational, the USBM's interest sprang from a realization that DSTP might provide the lowest impact alternative, particularly in the rugged terrain and high rainfall environment of Alaska's southern coast.

Although the USBM Alaska Field Operation Center literature search was extensive (containing 1,453 total references) and is certainly worth review for those keenly interested in DSTP, USBM officers found that much of the technology was published only in corporate reports or in regulatory agency documents (Baer, Sherman, and Plumb 1992).

TABLE 1.1 International DSTP sites currently operating, currently proposed, recently closed, and proposed but not implemented

DSTP Site, Mine Owner	Current Status	Solids Throughput (tonnes per day)	Nature of Process Plant
Currently operating			
Misima, Papua New Guinea	1998–present	20,000	Gold mill, cyanide, autoclave
Cayeli Bakir, Turkey	1993–present	2,000	Copper-zinc mill, flotation
Lihir Gold, Papua New Guinea	1996–present	3,500	Gold mill, cyanide, and cooling waters
Minahasa, N. Sulawesi, Indonesia	1996–present	3,000	Gold mill, roast, cyanide
Batu Hijau, Sumbawa, Indonesia	1999–present	160,000	Copper-gold mill
Huasco Iron, Huasco, Chile	1994–present	3,000	Iron ore pelletizing
Currently proposed			
Kensington Mine, Alaska	Under consideration	4,000	Gold mill
Awak Mas, S. Sulawesi, Indonesia	Under consideration	3,000	Gold
Petaquilla, Panama	Under consideration	90,000	Copper-gold
Simberi, Papua New Guinea	Under consideration	3,000	Gold
Ramu, Papua New Guinea	Under consideration	14,000	Ni/Co laterite autoclave leach
Tampakan, Philippines	Under consideration	50,000	Copper-gold
Moneo Nickel, New Caledonia	Under consideration	5,500	Ni/Co laterite autoclave leach
Namosi, Fiji	Under consideration	~100,000	Copper-gold
Recently closed			
Island Copper Mine, BC, Canada	Operated 1971–1995	30,000–60,000	Cu-Mo-Au flotation
Kitsault Molybdenum, BC, Canada	Operated 1980–1982	20,000	Molybdenite flotation
Black Angel, West Greenland	Operated 1973–1986	2,150	Pb-Zn flotation in seawater
Atlas, Cebu Island, Philippines	Operated 1971–1994	70,000–100,000	Cu flotation
DSTP proposed but not implemented			
A-J, Alaska	Plans to rejuvenate on hold	13,500	Gold-cyanide tailings
Voisey Bay, Newfoundland, Canada	DSTP evaluated as alternative	7,000–15,000	Nickel sulfide flotation
Quartz Hill, Alaska panhandle	On hold because of economics	80,000	Molybdenum sulfide flotation
Toquepala and Cuajone, Peru		100,000	Copper flotation

To assist in overcoming this obstacle, the USBM enlisted the assistance of several specialists in this particular field (several of these specialists are authors of chapters in this book) to compile information on DSTP. This resulted in the publication of two USBM volumes in 1993 and 1994—*Case Studies of Submarine Tailings Disposal: Volume I—North American Examples* (Poling and Ellis 1993), and *Case Studies of Submarine Tailings Disposal: Volume II—Worldwide Case Histories and Screening Criteria* (Ellis, Poling, and Pelletier 1994). Much of the information from these two USBM volumes was later incorporated into a special issue of *Marine Georesources and Geotechnology* titled "Submarine Tailings Disposal" (Ellis and Poling 1995).

Although U.S. Environmental Protection Agency (EPA) regulations appear to prohibit DSTP in most circumstances, regulations do appear to be sufficiently flexible to ensure that the alternative with the least impact can be permitted if an overpowering case is made. One key case study in this regard is the proposed Quartz Hill Molybdenum Mining Project in southeastern Alaska. At the proposed Quartz Hill Mine, after evaluating several alternatives, U.S. Forest Service managers identified DSTP as the preferred alternative for this particular operation, which was to be located 70 km east of Ketchikan, Alaska. U.S. Borax discovered this orebody in 1974, before the claims were encircled by the Misty Fjords National Monument and designated as a wilderness site in 1980. This wilderness monument specifically excluded 61,760 ha of central land for the Quartz Hill Mine project.

After U.S. Borax invested more than $20 million (of more than $100 million spent on the project as a whole) gathering baseline environmental data and funding four separate environmental impact statements (EIS), the EPA in 1990 denied application for a permit to discharge tailings at depth into Wilson Arm/Smeaton Bay but indicated a preferred site for marine tailing placement in an adjacent fjord, Boca de Quadra. This permit denial was never appealed, probably because of a falling demand for molybdenum and a decline of the price for this metal, which made the economics of the Quartz Hill Mine suspect. In 1991 U.S. Borax sold the property to Cominco and, to the authors' knowledge, it has not progressed further on plans to develop a mine at the Quartz Hill site.

Since the denial of the Quartz Hill permit in 1990, at least two other proposed new mines in southern Alaska have pursued plans to use DSTP; these are also listed in Table 1.1 as the A-J Mine and the Kensington Mine. It seems, then, that although very significant permitting difficulties certainly remain, mine developers are convinced that U.S. regulatory agencies can approve DSTP permits if a sufficiently compelling case is made.

Table 1.1 shows that island-based mine sites currently dominate the existing and proposed DSTP projects. Of these, Batu Hijau in Sumbawa, Indonesia (1,520 km east of Jakarta), is the most recent and by far the largest operation in terms of tonnage. This mine is currently in a ramp-up phase until probably 2002. Although the DSTP system is said to be operating well, very little detail has yet been published on this operation.

BRIEF DESCRIPTION OF THE DEEP SEA TAILING PLACEMENT TECHNOLOGY

DSTP is potentially applicable to mines located within about 200 km of a marine coast. Mill tailings composed of solid particulates and aqueous phase can sometimes be transported as slurry in a pipeline up to 200 km or more. Compared to the capital and operating costs of on-land impoundments, this type of slurry transportation can be economical.

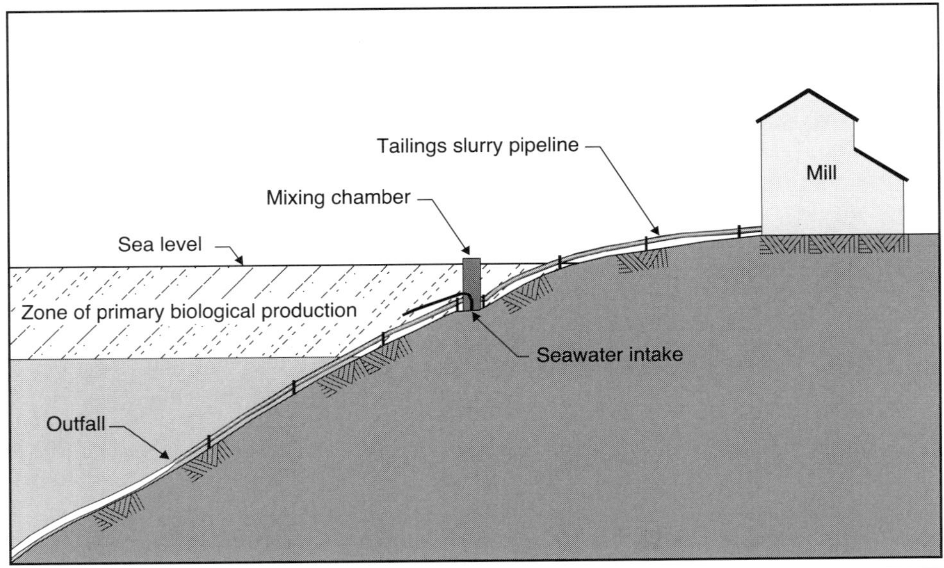

FIGURE 1.1 Schematic of a gravity-based typical DSTP system

In most locations, found much closer to the coast, gravity-fed pipeline systems, such as that shown schematically in Figure 1.1, can be employed. This system incorporates a seawater mix tank near the shore. The main functions of the mix tank are to serve as a deaerator (so that residual sulfide particulate will not be "floated" to the surface) and to eliminate the tendency of the lower density fresh water carrier liquid to become buoyant or rise as a plume toward the surface near the actual outfall or pipe terminus. A buoyant plume might carry fine particulate into the sunlit upper reaches of the seawater column, cause turbidity, and reduce photosynthetic processes. The zone of primary biological production in the ocean is normally within the top 40–80 m of the surface.

In some systems, particularly in lengthy submarine pipelines, the tailing slurry must be pumped because a gravity "head" can be insufficient to overcome pipe friction. In pumped systems, seawater dilution might not be needed to ensure that buoyant plumes will not occur. Rapid mixing of the fresh water carrier fluid with seawater beyond the outfall and discharge well below a pycnocline (or density gradient) in the ocean can both prevent a rising plume.

Beyond the outfall, the tailing slurry should flow to its eventual place of sedimentation as a density current or turbidity current. Ideally, suitable discharge sites should possess sufficient seafloor slopes and basin capacity for the life of the mine project. Otherwise, the project plan might have to incorporate the ability to move the actual pipe terminus periodically during the life of the project. One particular system (Minahasa; see Table 1.1) was designed so that the pipe terminus could be buried under a building mound of tailings kept fluidized, almost like a miniature submarine volcano, by continual pumping of either tailings or seawater itself.

The tailings slurry transport and eventual sedimentation characteristics in seawater depend critically on the solid particle size distribution, the densities of the component mineral particles, the particle shapes, and the slurry rheology. Obviously, the coarser and denser the particulate phase, the higher the carrier fluid velocity required for full suspension transport. Typically, finely ground copper mill flotation tailing solids (of which 60% might be finer than 74 µm) might require a slurry flow velocity of 2.0 m/sec for fully suspended transport in a pipeline. Tailings from pressure acid leaching of nickel laterite ores are typically much finer, perhaps 60% finer than 5 µm. Required transport velocity of 1.5 m/sec might suffice for such a slurry. On the other hand, very coarse (and dense) iron ore tailings might require pipeline transport velocities of 3, 4, or 5 m/sec. Upon discharge on a sloping ocean floor, fine tailing slurries can flow long distances as a coherent density current down a slope of only 1 or 2 degrees. Coarse, denser solids would settle out quickly on such a slope and build up a mound or deltaic type of deposit.

The emergent submarine density currents do not flow along the seafloor slope entirely coherently. In fact some natural "sorting" occurs, with coarser, denser particles dropping out as sediment closer to the outfall than the finer sized particles.

Characteristics of tailings that affect flow and sedimentation behaviors also have an impact on the stability of the tailings density currents and eventual sediment deposits. Ultrafine tailings particulates (<1–10 µm) can erode readily from deposits and become resuspended to move with ocean currents. Ultrafine components can also segregate from flowing density currents and might be carried upward with buoyant water plumes or be dispersed horizontally at pycnoclines. Fortunately, seawater, because of its high dissolved salt content, exerts a natural coagulating action on fine particles. Ultrafine components of tailings tend to adhere to the coarser fractions and flow and settle with them because of this coagulating action (Poling 1973).

Marine and submarine structures associated with DSTP systems must be secure and have a near-zero risk of physical failure. Tailing slurries are often highly abrasive, and both on-land and submarine components must be designed to resist wear failures. Seawater mix tanks are usually partially submerged and must be secure against wear failure but also resist any storm or wave damage. Subsea pipeline segments are heavily weighted and even sometimes buried through the intertidal zone to be secure on or below the sea bottom. Installations should include provisions for regular inspections to improve security. With the deepest DSTP system discharging at 385 m, remotely operated vehicles (ROVs) are usually used for underwater inspection.

OVERVIEW OF BIOLOGICAL CONCERNS

To some extent, DSTP can be expected to result in increases of suspended sediments, dissolved trace metals, and residual milling chemicals, as well as in the generation of an inundating sediment footprint in the receiving seawaters and on the seafloor. For these reasons, DSTP can present the prospect of both acute and chronic toxicities, bioaccumulation of metals, and habitat alteration. These potential impacts might apply directly to benthos, but fish, pelagic fauna (including phytoplankton), zooplankton, and microbiota might also be affected adversely. Detailed and scientifically peer-reviewed laboratory and field studies addressing specific biological aspects of DSTP are few and far between. In 1994, Kline published a review of "potential biological consequences" of DSTP. Kline reviewed more than 250 relevant reports and papers and prepared a 50-page report for the Alaska Bureau of Mines. Kline concludes from his review that, assuming tailings have "negligible reactivity, no resuspension of tailings and disposal into a deep, soft bottomed basin that is not critical habitat for commercially important species, the only likely consequence to biota is smothering of benthos or relocation of epibenthos for the duration of the operation and recolonization period." The importance of benthic smothering depends on the species present, particularly harvested crabs, mollusks, and flatfish. If the previous assumptions are not met, reductions in organism fitness may result because of increased metals and sediment in the water, also possibly leading to restrictions in human consumption of these organisms. Even if the organisms accumulate metals, biomagnification is unlikely.

Kline also points out some shortfalls in biological-type DSTP research, such as the need for more information on residual milling reagents, and for better understanding of benthic smothering, habitat alteration, and biota fitness. He also indicates that lack of peer reviews of many reports weakens any claims of low to neglible biological impacts from DSTP.

Kline went on to complete a Ph.D. thesis titled "Biological Impacts and Recovery from Marine Disposal of Metal Mining Waste" at the University of Alaska in Fairbanks in May 1998 (Kline 1998). Kline's research involved field studies of benthos colonization on trays of tailings and reference sediment placed on the seafloor near Juneau, Alaska, over a 22-month period. Kline found that froth flotation tailings from the proposed Kensington Gold Mine were recolonized by macrofauna similar to the natural sediments that were also field tested. He concluded that recolonization of these tailings (as a potential sediment footprint deposit) would be controlled by natural recruitment processes. However, it should be noted that the Kensington flotation tailings Kline tested are probably a good example of nontoxic tailings. They are composed mainly of siliceous minerals with relatively low (<1%) sulfide and heavy metal contents. This recolonization study does, however, confirm results from the ICM site (which has tailing characteristics similar to those of the Kensington Mine) that in spite of the lower organic content and coarser particle size of the tailings, they recolonized similar to the natural sediment trays.

At an opposite extreme, the Black Angel Mine on the western Greenland coast mined and processed a massive lead/zinc sulfide deposit that contained significant seawater-soluble forms of these two metals. As indicated in Table 1.1, this mine operated from 1973 until 1986 with tailings being discharged at a depth of 33 m into the relatively small (4 km long by 80 m deep) Agfardlikavsa Fjord. This fjord was seasonally stratified and often nearly saturated with oxygen. Some biota in this fjord were contaminated with levels of both lead and zinc during the mine's operating life. Once active mining had stopped, a waste rock pile at the fjord shoreline was found to have contributed most of the metal contamination in the water column. In 1990 this acid-generating waste rock pile was excavated and placed on top of the tailing sediment below 60 m of water depth. Ongoing studies indicate that this remediation measure improved the seawater environment considerably. Today, regulators estimate that only 0.1 tonnes of lead and 5 tonnes of zinc leach annually from both the undisturbed tailings and the partially oxidized waste rock on the bottom of the fjord (Asmund and Johansen 1999).

The Black Angel Mine is often cited as the example of the worst environmental impacts from undersea placement of tailings. Although some of this blame might be justified, most of the metal contamination came from waste rock and even from aboveground dusting-type sources that were poorly managed. Thankfully, today "the tailings from the Black Angel Mine now have been resting safe on the bottom of the sea for 9 years without giving rise to any significant lead pollution" (Asmund 1999). It is probably correct to conclude that preliminary test procedures recommended today would have likely excluded the particular design selected for DSTP for the Black Angel Mine site (Ellis, Poling, and Baer 1995). Certainly, acid-generating waste rock from a massive sulfide mining operation should not be deposited both on and below a shoreline where tidal wetting and drying cycles can accelerate sulfide weathering, acid generation, and metal release.

COMPARISON BETWEEN DEEP SEA TAILING PLACEMENT AND ON-LAND IMPOUNDMENT

By far the most common method of storing tailings around the world is in on-land impoundments, where cross-valley dams utilize the natural capacity of depressions or perimeter diking is erected around flatland impoundments that then stand proud of the landscape. In some instances, either all or a portion of tailing solids can be stored in underground excavations, perhaps as backfill for ground support, or in mined-out open pits, when multiple pit surface mining is used. Tailings are also sometimes stored at the bottom of natural freshwater lakes. Storage under a manmade or natural water cover is one of the best ways to prevent weather oxidation of reactive minerals such as sulfides. Subaqueous storage in perpetuity is therefore the most secure method for preventing acid rock drainage with its consequent metal leaching and potential contamination of downstream surface or groundwaters. Manmade, on-land submergence of tailings can be relatively simple in regions where precipitation exceeds evaporation. On the other hand, if acid generation is not a problem, on-land storage is then best suited to arid environments where evaporation exceeds precipitation. In these regions, tailing impoundments dry out, and no significant water need be discharged to the environment.

DSTP can be an alternative to on-land impoundment, mainly in rugged coastal or island terrains where precipitation exceeds evaporation and excess water must be discharged. In 1982, Caldwell and Welsh published a recommended procedure for comparing and eventually selecting either DSTP or on-land impoundment for rugged, high-precipitation environments. These authors provide a series of tables and checklists to enable eventual numerical comparisons of these two technologies.

Table 1.2 gives a quick comparison of DSTP and on-land systems, as compiled by this author. Some of the advantages and disadvantages of DSTP are expanded on in the following two sections.

ADVANTAGES OF DEEP SEA TAILING PLACEMENT SYSTEMS

- Given particular tailing and marine conditions, DSTP can have the lowest environmental impact of several alternative tailing placement strategies. The tailing solids must not leach toxins to the water column, and residual reagents cannot be measurably toxic in the receiving environment. The tailing sediment footprint must not obliterate a highly biologically productive ocean floor. Coral reefs are particularly sensitive to tailings sedimentation and must be protected. Once discharged, tailing effluent must not disperse into the uppermost euphotic zone, where it might impair photosynthetic processes. Detailed knowledge of ocean bathymetry, currents, and water column structure is essential to develop realistic models predicting the transport and stable sedimentation of particulate solids on the ocean floor.
- Tailing solids, once deposited in a canyon or large depression on the ocean floor, are generally stable and not susceptible to catastrophic failure as are some on-land impoundments retained by dams or dikes. In on-land impoundments, such

TABLE 1.2 Comparison between submarine tailing placement (STP) and on-land placement of sites with high precipitation rates

Parameter	STP	On-Land Impoundment
Land area covered by tailing solids	Zero (in pipeline only)	Usually several square kilometers; higher on flat terrain and for poorly consolidating nickel laterite tailings solids; lower in rugged terrains
Seafloor area covered by tailing solids	Normally several square kilometers; can be smaller in steep canyons or depressions	Zero
Seafloor/land area uses/impacts	Inundation where deposition rate is high; choose low-productivity floor area	Total inundation of land usually in valley or depressions that are the most highly productive local areas
Risks of system failure	Storage tanks, mix tanks, and pipelines must be secure from leaks, overtoppings, land slides, etc.	Dams/dikes can fail during extreme weather or seismic events; consequences of failure often very severe; pipelines must be secure from leaks; erosion a common problem
Risks of water contamination	Must demonstrate that solids do not leach toxins; residual milling chemicals must be nontoxic to seawater; near-zero risk to surface or groundwater	Seepages to surface and groundwater must be nontoxic or treated to become nontoxic; excess supernatants discharged with or without treatment
Risks of airborne releases of dust	Zero	Can be low to high depending on tailing fineness and remediation measures adopted
Reclamation	Near zero apart from removing on land structures (i.e., pipeline); monitoring of ocean can continue for several years	Remove structures; install spillways; re-slope; revegetate; maintain integrity and monitor possibly for an indefinite period of time; tailings difficult to revegetate because of high ionic content and poor consolidation properties of nickel laterite tailings solids
Aesthetic features	Very low visibility and impact	Visual impacts can be moderate to high for rugged to flat terrain; in-pit placement results in very low impact.
Prospects of water recycle	Limited to 50%–60% maximum; "flowable slurry" must be maintained	Can be high; normally 70%–80%, limited by interstitial water trapped in sediment; thickening and filtering tailings can increase to ~90%
Prospects of tailings retreatment	Near zero; deposits deep so dredging to recover too costly	Not difficult; tailings easy to re-mine and process
Prospects of remediation if impacts exceed predictions	Difficult to achieve; inert top cover could be placed to improve "isolation"; natural cover takes 2–3 years in most sites	Can stabilize dams/dikes; can intercept and treat seepages
Capital costs	Normally low; exceptionally long pipelines with pumping stations might narrow this advantage but usually won't eliminate it	Normally high; often $10–$100 million or more
Operating costs	Normally low, particularly for gravity-fed systems Environment monitoring costs normally exceed "on-land" costs	Can be medium to high depending mainly on pumping needs and elevation differences between plant and pond

failures are sometimes initiated by earthquakes or floods brought on by torrential rains or caused by overtopping by wave actions from extreme windstorms. Although the engineering and science of on-land tailing impoundment structures has improved dramatically during the last 2 decades, failures of tailing dams do still occur, sometimes with loss of life and always with substantial property damage. In many tropical island countries, such as the Philippines or Indonesia, such failures have invariably inundated productive lands, mangroves, and coral reefs.

- Tailing solids that might weather or oxidize, or both, and that generate heavy metal contaminated "acidic rock drainage" in an aboveground, on-land impoundment will not oxidize when placed at depth on the ocean floor. The fact that even oxygen-saturated seawater contains only 1/30th as much oxygen as air, along with the very low rate of diffusion of oxygen in seawater, effectively prevents oxidation of the solid particles. After mine closure, covering the top of the tailings sediment with natural marine detritus further prevents any potential long-term metal release by promoting diagenetic processes, such as sulfide precipitation, within the sediment.

- Slurry pipeline transport technology has improved to the point where mines up to 200 km or more from a satisfactory marine placement site can be feasible. In general, costs for DSTP systems are less than on-land placement alternatives, both from capital and operating points of view.

- DSTP minimizes land use for any mining project. This can be particularly important to island countries where valley and plains areas are often the most productive for farming and ranching.

- The salinity of seawater acts as a natural coagulant and causes fine-grained particles, such as tailings solids, to aggregate and settle much more quickly than in freshwater. Furthermore, seawater is naturally alkaline (~pH 8), which inhibits the solubilization of most metals.

- DSTP largely eliminates the risk of contamination of freshwater streams or lakes downstream from a mining project. Although low when modern technology is used, such risks are real. Contamination can occur as a result of seepages to either surface or groundwaters or direct discharge of supernatant waters from the top of tailing impoundments.

- Mine closures for DSTP systems often involve only physical removal of the pipelines and ongoing environmental monitoring. Closure of on-land tailing impoundments often involves construction of spillways, water treatment plants, considerable revegetation, and continuous monitoring of dam stability and water discharge quality. In some instances, risks of catastrophic failure of impoundment structures can persist.

- Because a DSTP system is never at the ocean surface, aesthetics are not an issue.

DISADVANTAGES OF DEEP SEA TAILING PLACEMENT SYSTEMS

- Even well-engineered DSTP systems can have some impact on the marine, physical, and biological ecosystems. Certainly, some benthic organisms will either move or be obliterated on areas of the seafloor subject to heavy solids deposition. It is imperative, then, to select low-value bioresource areas for placement of the tailings, and to minimize the deposition footprints in the area. Detailed oceanographic and engineering studies must be conducted to provide both regulators and the public assurance of this.

- Tailing solids that contain readily leachable (in seawater) toxins are not suitable for DSTP. The solids must be demonstrated to be relatively inert to seawater leaching, especially after deposition on the seabed. In addition, the liquid phase of the tailings must not contain toxic metal levels or residual milling chemicals that might be toxic.

- Tailing placement structures must be designed to withstand natural extreme weather events, seismic events, the dragging of ship anchors, and the like. Pipe wall thickness must consider abrasive wear of the slurry during the engineering design phase. All these elements require advanced engineering, construction, and operation technologies.

- Using current technology, if monitoring shows that on-land dams are becoming unstable, remedial measures can often be taken. Dams can be raised to increase freeboard or capacity, or both. DSTP systems often place the tailing solids under hundreds of meters of seawater covers. Once placed, dredging to recover is practically impossible.

- Not as much water content of the tailing slurry can be recovered for recycle back to the process plant. In closed-circuit on-land impoundments, as much as 80% of process water requirements can be recycled. This reduces a mine's demand for fresh make-up water and can conserve a resource that is sometimes scarce in arid climates. When using DSTP systems, tailing thickeners can often recover 50% of the contained water for recycle before discharging a thickened tailing.

- Environmental baseline and ongoing monitoring costs and time requirements to obtain necessary data for permitting may exceed similar costs for an on-land disposal system because specialized oceanographic vessels and remote sampling techniques are needed.

CONCLUSIONS

DSTP can sometimes have the least environmental impact of any available placement technology, although it is certainly no panacea. For new mines located close to a marine coast (within approximately 200 km) with deep water close to shore, the authors recommend a two-tiered evaluation procedure. The first phase can be completed within a few months, minimizing risk capital. The second, more detailed evaluation of DSTP will generally take more than 1 year so that complete seasons of environmental baseline oceanographic data can be acquired. The second phase study will be undertaken only if the first phase study shows no fatal flaws and the proponent is convinced that permits can be obtained in a reasonable time at a reasonable cost.

Deep Sea Tailings Placement Criteria and Recommended Procedures

Phase I

- The proposed mine and process plant must be within 200 km of a coast with deep water (>100 m) close to shore.

- Pipeline transport of tailings to the ocean is technically feasible; terrain must not involve too high a static lift for pumping; access to a pipeline right-of-way or corridor must be practically feasible.

- Bathymetric charts of the region must have reasonable resolution and accuracy or some preliminary oceanographic surveying must be done. The prospective tailing sediment footprint in either a basin or a canyon must have sufficient volumetric storage capacity to receive the entire volume of tailings sediment over the life of the mine. If the discharged tailing slurry flow is over a steeply sloping shelf to eventually sediment at great depth, volumetric capacity is not a prime concern.

- Subsea currents in the region of planned deposition should not be capable of resuspending tailings or of upwelling tailings into the euphotic zone or in toward the shore.

- The proposed tailings sediment footprint should be in a relatively low-value bioresource zone. A soft, muddy-type bottom is preferred. The footprint should be away from a commercial bottom fishery (such as a fishery for crabs, prawns, or flatfish). Sensitive areas such as coral reefs and mangrove forests must not be threatened by tailings sedimentation.

- Testing of tailing solids must show them to be relatively inert and not leach toxins to the seawater. Testing should also demonstrate that the aqueous component of the tailing slurry should not be acutely toxic at the pipe terminus. A seawater dilution zone should not be necessary to demonstrate that the tailings are nontoxic. Although this recommendation is contrary to some existing permits, it seems to be essential if DSTP is to justify claims of being capable of having the least environmental impact.

- Coastal installations such as pipelines, mix tanks, and possibly pump houses must be secure against all anticipated storm damage (including tsunamis) or seismic events.

- As early as possible, corporate plans should be effectively communicated, not only to regulators but also to the local community.

- Political, ecological, health, and social risks should be low. Prospects of achieving the lowest risk technology without undue prejudice must be high.

- Tailings transport and subsea flow and eventual deposition should be mathematically modeled, although only in a preliminary fashion at this phase. It is unlikely that sufficient data will be available for detailed modeling at this stage.

- Preliminary capital and operating costs should be estimated.

Once Phase I is complete, comparisons can typically be made between tailing placement alternatives.

Phase II. Successful completion of Phase I, with good prospects of achieving a reasonable permit, can mean that a Phase II evaluation might reasonably be initiated. The necessary aspects include:

- Detailed characterization of the tailing slurry—tonnages; volumes; densities; particle size distributions and ranges expected; detailed mineralogies; rheology as a pipeline slurry and as a density current flow along the seafloor; geochemistry; and toxicity tests for solids (leachability in seawater), liquids, and any residual milling chemicals

- Detailed characterization of the subsea receiving environment and variations expected with time (seasons, storms, etc.)—wave and current data, existing biosystems classifications and inventories, water quality data, and sea bottom characteristics

- Realistic/reliable modeling of tailings behaviors in the sea and as sediment on the sea floor—details of tailings transport and sedimentation footprints, estimates of turbidities within the water column, and predicted impacts on biosystems

- Monitoring programs to assess system performance and impact in terms of biodiversity and resource recovery—should be designed and implemented at least 1 year before operations begin and continued for several years after the mine is closed

- Socioeconomic assessments, examinations of potential reparations, and plans for ongoing community relations building

- Preparation of a detailed EIS

- Permitting of DSTP project—begin the permitting process as early as possible

Timing. Although DSTP is often a lower cost alternative (both capital and operating) than on-land impoundment systems, proponents must expect that achieving permits will take longer. This can result partly from the longer period required to acquire representative baseline data, but it can also result from the heightened sensitivities that can be expected from both regulatory agencies and the public. DSTP should be seriously pursued only if a proponent is truly convinced that a very low environmental impact can be achieved.

REFERENCES

Asmund, G. 1999. Personal communication.

Asmund, G., and P. Johansen. 1999. Short and long term environmental effects of marine tailings and waste rock disposal from a lead/zinc mine in Greenland. *Proceedings of the International Congress on Mine, Water and Environment for the 21st Century.* Seville, Spain.

Baer, R.L., B.E. Sherman, and P.D. Plumb. 1992. *Submarine Disposal of Mill Tailings from On-Land Sources—An Overview and Bibliography.* U.S. Bureau of Mines Open-File Report. Washington, DC: U.S. Department of the Interior, Bureau of Mines. 89–92.

Caldwell, J.A., and J.D. Welsh. 1982. Tailings disposal in rugged, high precipitation environments. An overview and comparative assessment. *Marine Tailings Disposal.* Ann Arbor, MI: Ann Arbor Science. 5–62.

Das, D.K., and G.W. Poling. 2000. Deep sea tailing placement systems. *Asian Journal of Mining.* February/March:31–34.

Ellis, D.V. 1989. *Environments at Risk.* New York: Springer-Verlag. 329 pp.

Ellis, D.V., and G.W. Poling, eds. 1995. Special issue: submarine tailings disposal. *Marine Georesources and Geotechnology* 13(1–2).

Ellis, D.V., G.W. Poling, and R.L. Baer. 1995. Submarine tailing disposal (STD) for mines: an introduction. *Marine Georesources and Geotechnology* 13(1–2):3–18.

Ellis, D.V., G.W. Poling, and C.A. Pelletier. 1994. *Case Studies of Submarine Tailings Disposal: Volume II—Worldwide Case Histories and Screening Criteria.* U.S. Bureau of Mines Open-File Report. Washington, DC: U.S. Department of the Interior, Bureau of Mines. 37–94.

——. 1995. Potential for retrofitting STD. *Marine Georesources and Geotechnology* 13(1–2): 201–233.

Kline, E.R. 1994. *Potential Biological Consequences of Submarine Mine Tailing Disposal. A Literature Analysis.* U.S. Bureau of Mines Open-File Report. Washington, DC: U.S. Department of the Interior, Bureau of Mines. 36–94.

——. 1998. Biological impacts and recovery from marine disposal of metal mining waste. Ph.D. thesis, University of Alaska, Fairbanks.

Poling, G.W. 1973. Sedimentation of mill tailings in freshwater and in seawater. *CIM Bulletin* 66:97–102.

Poling, G.W., and D.V. Ellis. 1993. *Case Studies of Submarine Tailings Disposal: Volume I—North American Examples.* U.S. Bureau of Mines Open-File Report. Washington, DC: U.S. Department of the Interior, Bureau of Mines. 89–93.

CHAPTER 2

Selection of Subsea Tailing Placement

George W. Poling

EXPLORATION AND DISCOVERY

The first real mining at the northern end of Vancouver Island began near Fort Rupert in 1849. Welsh and Scottish miners mined the area to supply coal fuel for Royal Navy and Hudson's Bay Company steamships (see Figure 2.1 for a map). Before that date, these two companies had bought some coal directly from the west coast aboriginals who undoubtedly had discovered the coal fields. Iron ore mining came next with significant production from the Benson Lake mine until 1967. Exploration for iron ore brought the Utah Construction and Mining Company (Utah Mining) to the north Vancouver Island area in 1949. Eventually this company established the ICM on the north shore of Rupert Inlet in 1971. Yreka Mine, located on the west side of Neroutsos Inlet (see Figure 2.1), was the first copper mine brought into production in this area. This mine operated off and on from 1902 until 1967 and disposed of tailings into Neroutsos Inlet (Waldichuk 1956), which is immediately south of Rupert Inlet (see Figure 2.1). Waldichuk, a world-renowned federal government "fisheries" scientist, found only minor impacts from this subsea tailing discharge to Neroutsos Inlet.

Prospector Gordon Milbourne originally staked the ICM property over the period from 1963 to 1965. In January of 1966, Utah Mining signed an option agreement on Milbourne's 150 claims. Although surface showings of copper sulfides had been found in the neighborhood, federal government geophysical maps showing a magnetic anomaly to the north of Rupert Inlet drew Milbourne to the initial staking activity in 1963. Utah Mining geologists ended up drilling a total of 128 diamond drill core holes (totaling more than 35,593 m) and sinking an exploration shaft 1.8 m by 3.7 m to a depth of 59 m with a 305-m-long cross drift at the bottom to prove up the Island Copper orebody. This exploration delineated a copper-molybdenum orebody made up of more

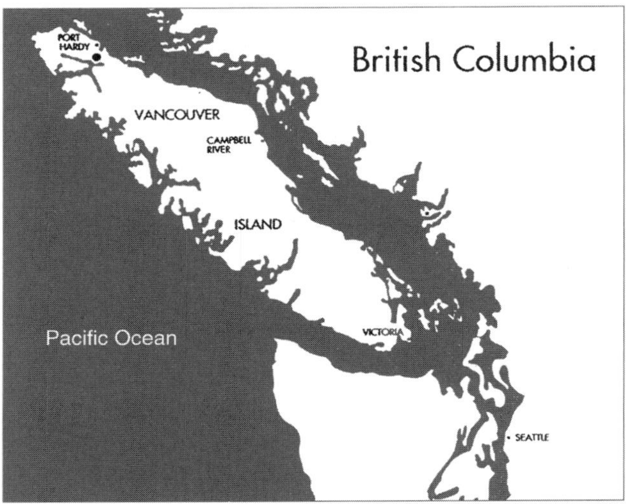

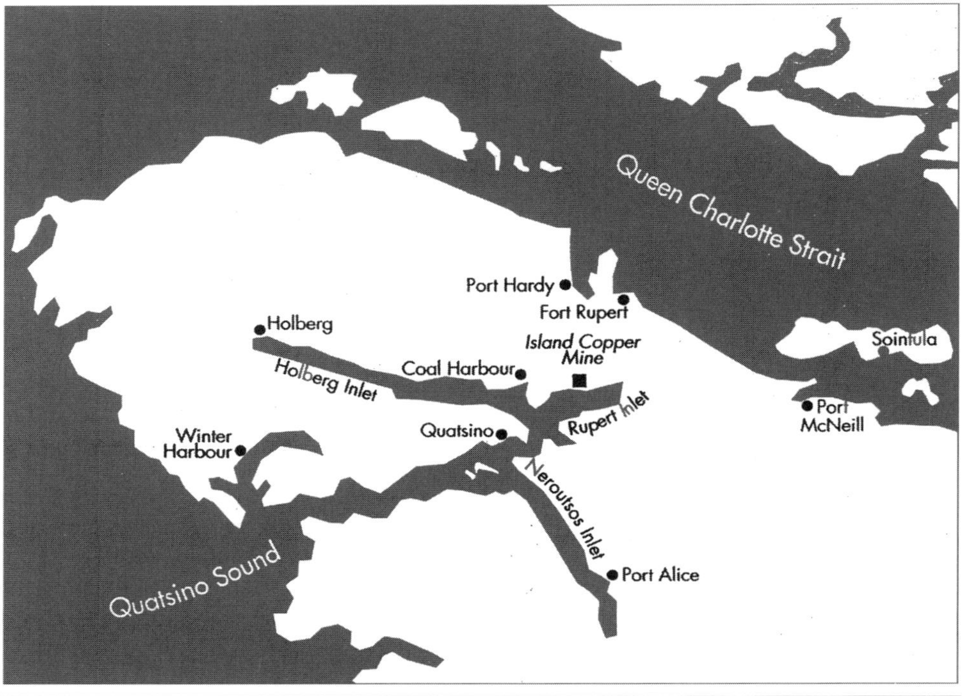

FIGURE 2.1 Location of ICM along the north shore of Rupert Inlet

than 254 million tonnes of ore containing an average of 0.52% copper plus 0.017% molybdenum metals. In June of 1969, Utah Mining announced that the ICM development was a "go" (*North Island Gazette* 1969). The mine was to be an open-pit truck and shovel-type mine feeding a mineral processing plant 30,000 tonnes of low-grade ore each day. The ore was to be crushed and ground to a fine particle size range approximately 60% finer than 74 μm before physically separating the solid particles via froth flotation into separate copper (as chalcopyrite) and molybdenum (as molybdenite) concentrates and tailings. Processing 30,000 tonnes per day of ore was projected to produce 630 tonnes of copper concentrate, 4.93 tonnes of molybdenite concentrate, and 29,365 tonnes of tailing solids particulates. Much of this history has been taken from *The Story of Island Copper* (Aspinall 1995).

PERMITTING THE MINE

Shortly after the exploration program began to indicate that prospects for developing a large-scale open-pit copper mine were good, Utah Mining began investigating tailing disposal alternatives at the site (in April 1968). Early work included bathymetric surveys of Rupert Inlet, evaluation of data from Canada's Federal Department of Fisheries (now Department of Fisheries and Oceans), bioassays of mill tailing effluent from a pilot-plant operation, design of engineering structures, evaluation of the probable effects of DSTP on marine life, and visits to other mine sites employing marine tailing placement. Potential on-land sites were evaluated with the knowledge that high levels of precipitation (~1.8 m/yr) would necessitate discharge of an impoundment supernatant to the inlets. In addition, the probability of seismic accelerations exceeding 0.1 × gravity was high at this site. For these reasons, construction of secure impounding dams would be more difficult. On the basis of these evaluations, coupled with close consultations with both Province of British Columbia Mine and Environmental regulators and Federal Department of Fisheries officers, Utah Mining concluded that DSTP was in fact the preferred option for the ICM site. On October 2, 1969, the company applied to the then-Provincial Pollution Control Branch for a permit allowing the discharge of thickened ICM tailings at a pipe terminus depth of −50 m in Rupert Inlet. The tailings were projected to flow into the deepest portions of the inlet (originally ~153 m at the deepest point) and eventually fill approximately one-tenth of the volume of the inlet with tailing sediment over the proposed 25-year life of the mine.

Publication of the application for a permit in October 1969 rapidly focused both public and media attention on Utah Mining's proposal. This was a period of rapidly escalating public concerns about environmental protection. William Venables, then director of the Pollution Control Branch, received approximately 140 letters objecting to the application. After replying to all letters with more information, Venables received follow-up replies from four objectors. He then decided to hold a public hearing in Port Hardy on December 2, 1970, to receive briefs from these four respondents, government scientists and the proponent, Utah Mining. The 12-hour hearing was well attended and drew considerable media attention.

PREOPERATIONAL BASELINE DATA

In April 1968, Utah Mining began to investigate the feasibility of placing mill tailings at depth into Rupert Inlet. Early work included:

- Depth surveys of Rupert Inlet
- Evaluation of data on Rupert and Holberg Inlets compiled by the Federal Department of Fisheries

- Chemical and bioassay tests of effluents generated by a pilot-plant metallurgical test operation
- Study of settling characteristics of the projected mill tailings
- Preliminary engineering design of a DSTP system
- Valuations of proposed placement impacts on aquatic life in the inlet

After becoming convinced that DSTP was preferable and filing an application for the permit on October 2, 1969, additional preoperational studies were conducted, including biological surveys of deep bottom fauna in January 1970 and again in October 1970, current surveys, temperature and salinity measurements, and sampling of bottom sediments. Results from all these studies were compiled in a brief for presentation by Utah Mining to the public hearing in Port Hardy (Utah Construction and Mining, Ltd. 1970).

The folowing are some of the more interesting monitoring results and controversies included in this brief:

- Tailing characteristics and bioassays
- Tailings quantity to be discharged: 42,300 m^3/day containing 40%–50% by weight of finely ground (−0.070 mm) solids
- Chemical compositions of tailing solids shown in Table 2.1

Effluent quality to be expected was based on representative samples from the liquid portion of the pilot-plant test work on the copper flotation circuit. Projected 1:1 dilution of effluent concentrations with seawater were given as shown in the second column of Table 2.2.

Table 2.2 includes permit limits not to be exceeded in the eventual permit (PE 379-P) for undiluted tailings. Based on the comparison shown in this table, it would seem that the provincial regulators simply took the numbers proposed by Utah Mining and cut them in half (which would incorrectly assume that the seawater diluent contained zero concentrations of each element). In 1971 these were extremely low levels of dissolved metals for a mining effluent permit. In fact, at that time, the metal detection limit for the best local commercial analytical laboratories was 0.01 ppm for most elements. Values less than that would simply be reported as <0.01 ppm. Many commercial laboratories at that time analyzed only down to 1 ppm.

TABLE 2.1 Chemical analyses of various size fractions of solid portion of effluent from flotation test of ICM ore

Sample Number	As	Cu	Co	Cr	Mn	Mo	Ni	Pb	Zn
					mg/kg (ppm)				
39136	0.31	89	<5.0	10	81	<1.0	5.0	24	39
39137	0.23	88	<5.0	9	76	<1.0	3.5	12	44
39138	0.59	71	<5.0	10	78	<1.0	6.0	11	36
39139	0.49	93	<5.0	30	80	<1.0	5.0	12	40
39140	0.31	89	<5.0	9	79	<1.0	3.0	8	37
39141	1.4	59	<5.0	4	78	<1.0	5.0	15	32

Taken from Utah Construction and Mining, Ltd. (1970).

TABLE 2.2 Projected 1:1 dilution effluent concentrations in Utah Construction and Mining Ltd. (1970) compared to eventual permit awarded

Element	ppm	Permit Concentrations (not to be excluded, ppm)
CN	0.087	0.05
Zn	0.0125	0.022
As	0.004	0.008
Mo	0.01	0.02
Cd	less than 0.025	0.05
Cr	less than 0.023	0.05
Cu	less than 0.0025	0.005
Pb	less than 0.0025	0.005
Co	less than 0.025	0.05
Ni	less than 0.0025	0.005

Biological and Chemical Nature of the Proposed Receiving Water Body

Rupert/Holberg Inlets and Quatsino Narrows were well recognized as important fishing waters. The Marble River discharges into the southwest end of Rupert Inlet and supports significant escapements of all five species of Pacific salmon as well as steelhead trout. In addition, an intermittent crab fishery existed in both Rupert and Holberg Inlets above a depth of 75 m. The deeper portions of these inlets were populated mostly (>70%) by burrowing worms. T.W. Beak Consultants conducted a field survey for Utah Mining between January 17 and January 21, 1970 (Utah Construction 1970). They established 21 biological sampling stations, with triplicate samples taken by Ponar dredge at each station. Water depths and water transparency (via Secchi disc) were also recorded at each station. Water column samples were taken at three depths (3 m, mid depth, and near bottom) at five "chemical sampling stations." On the basis of their survey, T.W. Beak concluded that:

- Benthic productivity in the deep central basin of Rupert and Holberg Inlets is not high.

- The deposition of tailings in this basin would reduce but not eliminate production of the predominant bottom-associated organisms.

- Providing the tailings settle as predicted, they should present no hazard to the benthic fauna of the more productive shallow peripheral areas.

- Results of chemical analysis were characteristic of uncontaminated coastal marine waters.

Physical Characteristics of the Rupert/Holberg Inlet System

Existing hydrographic surveys of Rupert/Holberg Inlets were reasonably accurate and detailed as shown in Figure 2.2 (taken directly from the 1970 Utah Mining brief). Depths shown in this figure are in feet (as in the original figure). The bottom trough of Rupert Inlet slopes west to achieve a maximum depth of >152 m (>500 ft) immediately north of Quatsino Narrows (Figure 2.2). Quatsino Narrows is a very narrow passage that has a shallow sill (<18 m) and leads to the Pacific Ocean via Quatsino Sound (see Figure 2.1).

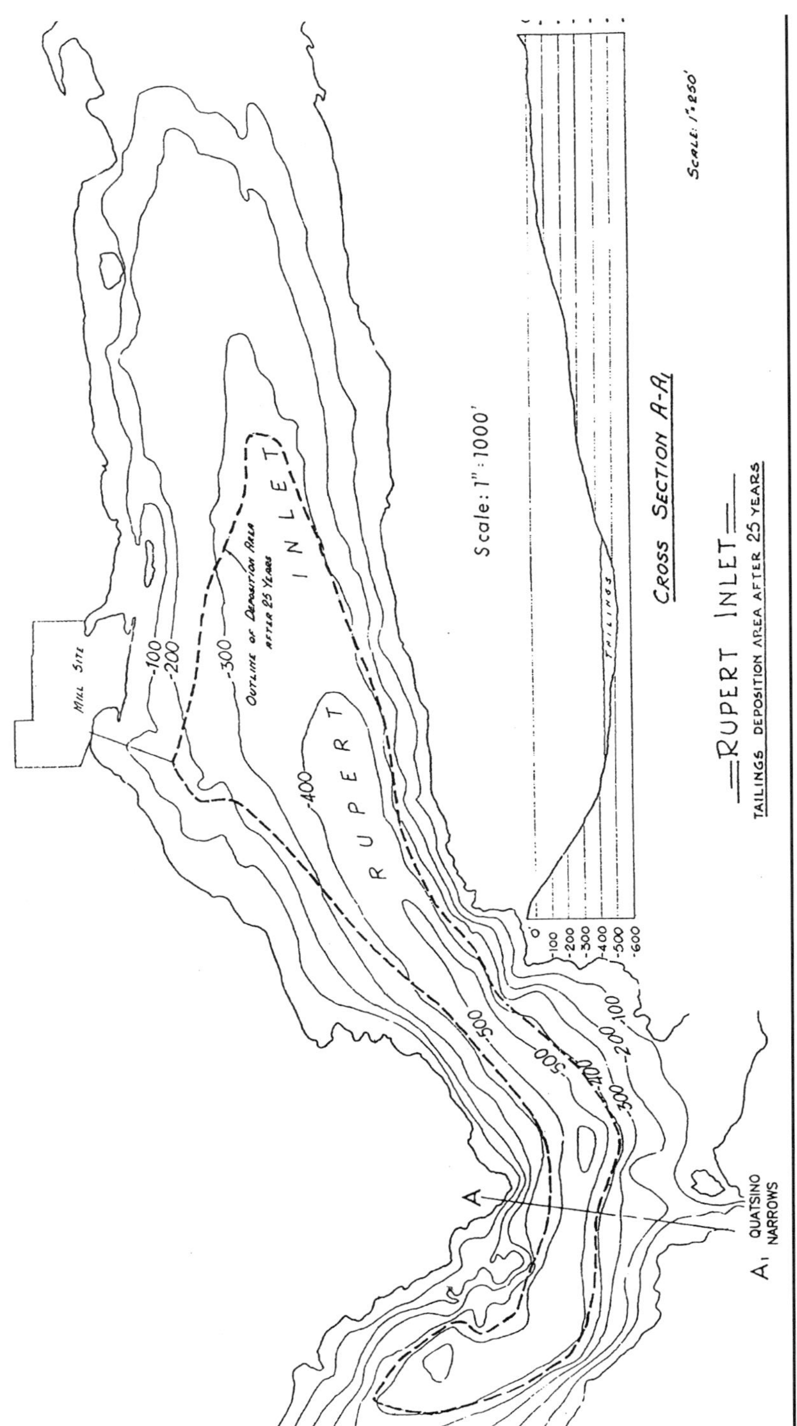

FIGURE 2.2   Projected area of tailing deposition in Rupert Inlet from 25 years of discharge

Source: Utah Construction and Mining, Ltd. 1970.

BC Research was retained to conduct current measurements at two locations in Rupert Inlet over the period May 19–22, 1970. One station ("B") was near the proposed outfall discharge site offshore from the mine (see Figure 2.2), and the second station ("A") was almost midway between the mouth to Quatsino Narrows and Hankin Point (which is at point A in Figure 2.2). A Savonius Q9 rotor current meter was used to measure both current speeds and directions at these two stations from a 20-m anchored boat. Currents at station B were all less than 0.18 m/sec, and good quality data were obtained at this site. Difficulties were encountered in taking reliable measurements at station A because of the strong tidal currents. The boat was sometimes dragged off station during flood tides, and the meter support cable was often far from being vertical. In hindsight, a moored meter was required (and has since been used at that location). BC Research recorded several relatively high velocities (up to 0.7 m/sec) both near surface and at depth.

One reading of 0.6 m/sec was recorded at 150 m (490 ft) and another of 0.62 m/sec at 107 m (350 ft).

At the time, Utah Mining employed two consulting engineering specialists to oversee and review the BC Research current data. Both appear to have rejected BC Research's relatively high current readings at depth as spurious and "questionable." J.W. Johnson concluded that "In general the current speeds near the bottom are relatively low," and "There is no evidence in the records that the high surface currents on the flood tide flowing through Quatsino Narrows 'dive' into deep water under the surface layers of Rupert Inlet" (Utah Mining and Construction Ltd. 1970).

Johnson's views were echoed by Ralf Carter (another consultant to Utah Mining), who concluded that "The incoming ocean water during the study period appeared to be mixed with inlet water and to layer above 300 ft depth (91 m) rather than maintain its identity and flow to the bottom in a density current" (Utah Mining and Construction Ltd. 1970). Carter's conclusions were based both on the BC Research results and on his own density survey conducted on June 17, 1970. The conclusions by Johnson and Carter were also based on the findings that sediment cores (from a 38-mm-diameter gravity core barrel sampler) collected on May 28, 1970, contained very cohesive silts and clays. Because they concluded that even cohesive silts and clays would remain only if bottom current speeds stayed below about 0.12 m/sec, low bottom currents were indicated. BC Research also obtained fine-grained cohesive sediments at station B (near the eventual outfall). They were unable to obtain sediment samples using a Phleger torpedo-type core sampler at station A (off Hankin Point).

Projected Effects of Tailing Discharge on Receiving Waters

Utah Mining predicted that the flotation tailings, thickened to contain approximately 50% solids by weight, would be diluted with seawater (at approximately 1:1 ratio) in a mix tank at the shoreline and then discharged at a depth of at least 46 m (150 ft). Bioassay tests on the effluent, conducted on pilot-plant- and laboratory-generated tailings, showed that the liquid fraction of the tailings was not acutely toxic to fish even at 100% concentration. Thus the mining company projected that the liquid components would not have a significant impact on the receiving waters.

The fine-particulate solids fraction of the tailings was projected to flow out the end of the outfall pipe, mix with more seawater, and flow downslope in Rupert Inlet as a somewhat cohesive density current. Utah Mining projected that "the solid fraction of the tailing will settle in the deeper portions of the inlet, generally below 91 m (300 ft) and that the natural water movements in the inlet are such that they will not disturb the settled solids. There will be a reduction in deep water fauna—worms, small clams and shrimp-like organisms. This reduction will not affect the production of significant food species, crab and salmon, since these occupy shallow areas and surface waters." Utah Mining further predicted that "the deposition area over the projected 25-year life of the mine is estimated to be about 648 ha (1,600 acres) occupying approximately one-tenth of the volume of the inlet." The dashed line in Figure 2.2 represents the projected outline of the deposition area after the 25-year life of the mine.

Author's note: As it turns out, the highly simplistic model predicting the eventual tailing footprint was amazingly accurate from an engineering point of view. Elaborate bathymetric and seismic surveys indicate that more than 99% of the solids fraction of the tailings ended up occupying a footprint nearly identical to that shown in Figure 2.2.

PUBLIC HEARING

Probably the most effective objector to the ICM permit application at this hearing was a then 23-year-old graduate student from the University of British Columbia (UBC) named Patrick Moore. Moore was born and raised in the Rupert Inlet region, and his Ph.D. thesis subject concerned alternative land uses at the ICM mine site. His brief, dated July 24, 1970, maintained that the inlet had continual vertical mixing contrary to the company's reports, which indicated a stratified water column in Rupert Inlet. Moore asserted that there was mixing of deep water with the surface water, which would bring to the surface some of the suspended tailing solids.

He also predicted that "at the very least all the bottom-dwelling organisms will be destroyed, at worst the inlet may be polluted beyond recovery" (Moore 1970). As an alternative, he proposed "the tailing should be continuously returned to the mine site. Here they should be used to fill in the hole from which they were removed." Although this would have been practically impossible unless the tailings were all temporarily stored in an on-land impoundment until after the open pit mine was completed, for safe storage in perpetuity, this alternative otherwise had considerable merit.

The well-publicized objections of Patrick Moore were probably his springboard to cofounding the Greenpeace organization shortly thereafter. He served as president of Greenpeace for many years and has been an active environmentalist essentially all of his life. As an aside, although his predications of vertical mixing were periodically correct, his predictions of adverse biological impact were not. Moore has now acknowledged that, in retrospect, DSTP at ICM had the lowest environmental impact. In a paper published on his "Greenspirit" Web site in 2000, Moore and coauthors C. Pelletier and I. Horne concluded: "It is apparent from the findings of the monitoring program that the mine tailings were dispersed more widely then [sic] was originally predicted. It is also true that widespread heavy metal contamination has not materialized. In retrospect it appears that submarine tailing disposal has resulted in far less severe environmental impacts than if the alternative of land disposal had been adopted" (Moore, Pelletier, and Horne 2000).

Although the public hearing conducted by the British Columbia Pollution Control Branch was controversial and attracted much public attention, Utah Mining received a permit enabling DSTP on January 20, 1971. Appendix A contains a copy of the initial permit (379-P). The permit set limits on physical and chemical parameters and biological nontoxicity of the tailing discharge. The company had to post a $1.5-million bond and assure construction of an emergency tailing impoundment with 6 months operating capacity. Utah Mining was also required to retain an independent scientific advisory committee to assist in establishing a comprehensive monitoring program, outlining procedures for sampling and establishing reliable analytical results, and preparing assessment reports for submission to the Pollution Control Branch. If the monitoring program indicated that unacceptable impacts were in fact occurring, the mine would have had to immediately divert tailings to the emergency tailing impoundment and then build alternate storage facilities.

The initial independent agency contract was awarded to UBC to administer the work of 15 university professors. Fields of specialization included physical and chemical oceanography, marine biology, ecology, chemistry, geology and sedimentology, mine engineering, and mineral process engineering. The university, in consultation with ICM staff and the British Columbia Pollution Control Branch, designed and implemented a comprehensive environmental monitoring program and several research programs to evaluate impacts of the tailings discharge on the marine ecosystem (Evans, Ellis, and Pelletier 1972). The university also assisted the mine's staff in establishing a state-of-the-science environmental sampling system and analytical laboratory at the mine site. Over the years the independent agency has comprised scientists and engineers from the three major universities in British Columbia—UBC, the University of Victoria, and Simon Fraser University—as well as from Rescan Environmental Services Ltd., a Vancouver-based consulting firm. Many graduate research theses have emanated from this monitoring contract with the university professors.

Members of the original independent agency included:
Department of Mining and Mineral Process Engineering, UBC

- Professor J.B. Evans
- Dr. J. Leja
- Dr. G.W. Poling

Institute of Oceanography, UBC

- Dr. G. Picard
- Dr. A. Lewis
- Dr. E. Grill
- Dr. M. Taylor

Department of Geological Sciences, UBC

- Dr. J. Murray
- Dr. K. Fletcher
- Dr. R. Chase

Department of Civil Engineering, UBC

- Dr. M. Quick

Department of Biology, University of Victoria (appointed 1972)

- Dr. D.V. Ellis

Dr. Glen Geen, a fisheries scientist from Simon Fraser University, also served with the advisory committee for several years before his untimely death.

In 1978, misinformation provided to the media led to headline charges that the independent monitoring agency was condoning the "killing of Rupert Inlet" by the tailing discharge. Within a week, nearly 50 papers across British Columbia featured this highly inaccurate story. UBC's president initiated a formal review of the charges, and the federal and provincial governments set up a joint team to evaluate and report on such charges. Eventually both the university's investigative lawyer and the joint-government "Waldichuk–Buchanan" report of 1980 debunked the media charges. In fact, the report complimented the comprehensive monitoring program and the results obtained, which showed very minimal impacts on the inlet ecosystem (Waldichuk and Buchanan 1980). The Waldichuk–Buchanan report concluded that the submarine placement system was performing such that "there was no basis for changing to any of the other alternatives for tailings disposal."

One outcome of the controversy was that the mining company commissioned an engineering firm (Ker, Priestman and Associates Ltd.) to prepare a "Study of Tailing Disposal on Land." Their report concluded that a land disposal site north and east of the present open pit could have been developed to satisfy mill needs for 18 years at an ore throughput rate of 34,000 tonnes per day (Ker, Priestman and Associates Ltd., 1975). The capital costs for building such an on-land alternative was then (1975) estimated to be $24,200,000. In addition, the operating cost of this on-land alternative would have been expensive, mainly because the abrasive tailing slurry would had to have been pumped up a static lift of 106 m. All in all, on-land tailing disposal costs would probably have approached $1.00 per tonne of ore processed.

Perhaps even more important than the partial sterilization of 133 ha of productive timbered and wildlife habitat land by the impoundment, the high precipitation and low evaporation rate at this site would have meant that surplus water on the pond would have been produced at a rate of 50,000 L/min. Although nearly half of this could have been recycled to the mill for process water, the other half would had to have been discharged to the environment. That effectively meant supernatant discharge to Rupert Inlet. From what we now know, these tailings might have become acid generating. Hence this pond supernatant would have likely required treatment before discharge for many years after mine closure. From several points of view (now in hindsight) the on-land alternative to the deep sea tailing placement system at the ICM was decidedly a second choice and not a preferred option.

Another outcome of the controversy was that the university chose to remove itself as contractor, recommending that the mining company contract directly with the professor advisors and independent consultants. The reorganized group, now reduced to four professors (Ellis, Parsons, Murray, and Poling) and C. Pelletier of Rescan Environmental Services Ltd., became the Technical Environmental Advisory Committee (TEAC). This committee has continued to serve in this advisory capacity to both the company and the Province of British Columbia until the present, and this group comprises the authors of this book. The current committee members, most of whom have served on this agency/committee for more than 30 years, meet at least semiannually with at least one visit per year to the mine site. This committee continues to provide the mine and regulatory agencies with an external evaluation of monitoring methods, procedures, and quality assurances and also supplies a critical review of each annual environmental assessment report (AEAR). Each member signs a letter attesting to the accuracy of data contained therein. The advisory committee members have committed to remain as advisors through the post-closure monitoring.

One unfortunate outgrowth of the change in the numbers and status of the TEAC was a subsequent decline in the numbers of post-graduate thesis students working on the ICM DSTP project. In the early years of this mine's life, graduate thesis work focused more on the physical and geomorphological aspects of DSTP. Gradually, university research at this site changed to emphasize environmental and biological topics. Student projects were often funded jointly by monies from both BHP Billiton (formerly Utah Mining) and government sources. Most of this research work has been published in internationally refereed technical journals and proceedings of conferences and books.

Field data collected at the ICM from 1970 to 1996 are available at <www.westlink.com>. An inventory and a bibliography can also be accessed at <http://gateway3.uvic.ca/archives/featured_collections/mesc/home.html> or by sending an e-mail to dvellis@uvic.ca.

MONITORING OF ENVIRONMENTAL IMPACTS

The Province of British Columbia Pollution Control Permit No. 379-P (shown in Appendix A) specified parameter limits and monitoring requirements. At that time the nominal capacity of this mine and processing plant was 30,000 tonnes of ore processed per day. This initial permit specified that slurry effluent to be discharged to Rupert Inlet had to be less than 42 million L/day (9.3 million gal [Imp]/day) at less than 50% solids by weight. A host of parameter limits was also specified for compliance. For example, dissolved copper had to be less than 0.005 ppm; dissolved nickel, less than 0.005 ppm; dissolved zinc, less than 0.022 ppm; and dissolved molybdenum, less than 0.02 ppm. In addition, the environmental monitoring program summarized in Table 2.3 was undertaken at the frequencies specified and under the direction of the independent agency. The agency (contracted to UBC) also had the assignment of assessing data collected and issuing AEARs. Their initial report, *Summary Report, Pre-Operational Phase, March–September 1971, Environmental Control Program, Island Copper Mine, Rupert Inlet, BC*, was issued in January 1972 (ICM 1972). The agency's report on the initial production year, October 1971 to September 1972, was issued in July of the following year (ICM 1973). This initial summary report was divided into three sections: Physical Parameters, Chemical Parameters, and Biological Parameters. This and subsequent annual reports were often augmented by appendices that described additional studies.

Here are some of the conclusions drawn from monitoring discharge of approximately 6 million tonnes of tailing solids during this first production year:

- Tailing solids deposited mainly in the Rupert Inlet trough bottom achieved a thickness approaching 10 m in close proximity to the pipe terminus.

- Where tailing deposits exceeded 50 cm thickness, benthos appeared to be obliterated.

- Transmissometer surveys indicated that tailings-generated turbidity within the seawater column stayed below the euphotic zone.

- Phytoplankton and zooplankton populations in Rupert/Holberg Inlets appeared unaffected by the tailing discharge.

- The tailing discharge had no measurable effect on measurements of temperature, salinity, pH, dissolved oxygen, total cyanide, and mercury or on dissolved arsenic, cadmium, chromium, cobalt, copper, iron, lead, manganese, molybdenum, nickel, or zinc in the seawater column.

The initial production year report also found that there was no conclusive evidence to suggest that the mine tailing discharge was adversely affecting the Rupert/Holberg/Quatsino Sound ecosystem.

TABLE 2.3 Outline of environmental control monitoring program, ICM third production year, October 1973 to October 1974

	Description	Frequency	Objective
Marine Program			
Seismic survey	Bottom profile and sediment distribution	Annually, October 1974	Record tailing distribution
Bottom coring	Cores at 24 stations—log and measure tailing thickness on bottom	Quarterly	Visually determine tailing distribution
Bottom grabs	Sediment sample for heavy-metal analysis	Annually, March 1974	Chemically delineate the spread of tailing
	Collect benthic samples at 24 stations—log, sort to polychaetes and others, count and weigh, prepare samples for long-term storage	Quarterly	Monitor benthic population changes
	At 4 stations collect benthic samples for species-diversity study	Quarterly	Monitor benthic population diversity
	Collect benthic samples at 24 stations—log and forward to consultant for detailed identification	Annually, October 1974	Monitor in detail change in benthic communities
Water column	At 7 stations profile temperature, turbidity, color, transparency, and suspended solids	Monthly	Record water-column physical properties
	At 7 stations profile salinity; alkalinity; pH; dissolved oxygen; spent sulfite; "total" As, CN, and Hg; and dissolved and particulate Cd, Co, Cr, Cu, Fe, Mo, Mn, Ni, Pb, and Zn	Quarterly	Record water-column chemical properties
	At 120 stations determine clarity of water by use of transmissometer	Annually, April 1974	Record clarity of water column
	At 7 stations sample for chlorophyll A standing crop	Monthly	Record standing crop of primary-producers in water column
	At 4 stations zooplankton samples collected by quantitative horizontal tows and vertical hauls; samples are sorted, counted and identified, and analyzed for heavy metals	Quarterly	Record abundance, diversity, and metal concentration of primary consumer in water column

(Table continues on next page.)

TABLE 2.3 Outline of environmental control monitoring program, ICM third production year, October 1973 to October 1974 (continued)

	Description	Frequency	Objective
Intertidal	From 16 plates estimate growth rate and sediment deposition	Monthly	Record primary production in intertidal area and estimate sediment deposition
	From 7 stations collect various species of intertidal invertebrates and intertidal fish; weigh, measure, and determine metal concentration	Quarterly	Record metal concentration in intertidal organisms
Fishing	At specific sites collect fish by various methods; identify, measure, weigh, sex, and determine metal concentration	Quarterly	Record metal concentration in fish and estimate population
Crabbing	At 6 stations collect crabs; identify, weigh, measure, sex, and determine metal concentration; male–female frequency distribution and population estimate	Quarterly	Record metal concentration in edible crabs
Fresh Water Program			
Water Sampling	At 9 midstream locations collect samples for temperature; pH; alkalinity; dissolved solids; suspended solids; turbidity; color; hardness; dissolved oxygen; sulfates; nitrates; total extractable Hg and As; and dissolved and particulate Fe, Cd, Cu, Co, Cr, Mo, Pb, Zn, Ni, and Mn	Quarterly	Record chemical characteristics of water flowing into inlet
Meteorological Program			
	At the mine site record temperature, wind, precipitation, cloud cover, and sea state	Hourly	Record meteorological conditions
Effluent Discharge Program			
	From weekly Composites of daily samples taken from the thickener underflow determine pH; % solids; temperature; total cyanide; total mercury; and dissolved Cu, Mo, Cd, Cr, Co, Fe, Pb, Mn, Ni, Zn, and As	Daily samples analyzed weekly	Monitor physical and chemical characteristics of effluent
	Samples of final effluent sent out for 96-hour median tolerance level bioassay tests	Biweekly	Record tailing toxicity
	Determine effluent volume	Continuous	Record volume of effluent discharged to sea
	Study composition of tailing, settling rates, and leaching potential	Ongoing research	Research activity

On occasion during this initial production year, the ICM effluent exceeded discharge permit limits (some of which were judged to be unrealistically stringent) with respect to allowable discharge volume (i.e., 45 million gal [205 × 10^{6}L] on some days versus the 42 million gal [191 × 10^{6}L] permitted); total mercury and arsenic contents; and dissolved copper, nickel, lead, zinc, manganese, and molybdenum contents. Estimates were provided that tailings-derived dissolved metal inflows into Rupert Inlet were a small portion of natural inflows (with the exception of molybdenum, which amounted to ~56%). With tailing effluent inflows averaging 33 million L/day, fresh water inflow averaged 72,000 million L/day, and average seawater exchange amounted to more than 100,000 million L/day.

Because no adverse effects resulting from the dissolved chemicals in the tailing discharge could be detected, the independent agency recommended to the director of the Pollution Control Branch that the existing permit be amended to conform to the recently promulgated "A level" chemical specifications in its *Pollution Control Objectives for Mining, Mine-Milling and Smelting Industries in B.C.,* dated March 1973 (Province of British Columbia 1973). This recommendation was accepted, and permit PE-379 was subsequently revised.

Monitoring of the impacts of the ICM tailing discharge into Rupert Inlet has continued to the present day. The permit has been amended several times (September 21, 1971; December 6, 1971; April 17, 1972; June 30, 1977; and December 24, 1985) to accommodate both recommendations from the independent agency as well as requests from Utah Mining (now BHP Billiton) to accommodate increases in production. Appendix B contains a copy of the latest permit covering the ICM's production period.

PROBLEMS AND CONTROVERSIES DURING MINE LIFE

Early Problems with the Outfall System

In the originally installed DSTP system, thickened tailing slurry (at 35%–45% solids by weight) flowed by gravity down a 1-km-long × 0.86-m-diameter steel pipe into a seawater mix tank on the shoreline of north Rupert Inlet. The tailings flowed in this pipe at approximately 1.5 m/sec velocity and then poured into the top of an open-top cylindrical seawater mix tank ~3 m in diameter × 14 m tall and anchored to a massive concrete foundation. Taking static head off the tailing slurry by allowing it to "pour" into the tank through a slurry–air interface resulted in entrainment of air within this slurry on the order of 20% by volume. This mix tank was also equipped with a seawater intake line that drew seawater into the tank through a rock filter at ~6 m depth via a one-way valve. By design, seawater to dilute the slurry 1:1 was to flow by gravity into the tank because the projected higher average specific gravity (1.2) of the seawater-diluted slurry was to depress the slurry level ~3.5 m below the existing sea level (seawater has a specific gravity of 1.03). Air entrainment reduced the slurry density in the tank, which lowered the slurry level depression to less than 2 m, reducing the seawater dilution. The air entrainment also resulted in air bubbles emanating out of the pipe terminus at −50 m depth together with the tailings slurry. Some of the residual pyrite in the tailings became hydrophobic and floatable in this seawater-mixed slurry and floated to the surface of Rupert Inlet to create a metallic-looking "slick."

In February of 1974, a large steel "bubble hood" was installed above the subsea pipe terminus to catch these floating pyrite particles. The captured floating particles were carried back to shore via a 15-cm-diameter polyethylene pipe. At shore the sulfides were trapped within a floating boom system and periodically removed from the surface. Although this remedial system worked reasonably well, it would have presented a long-term maintenance problem.

The most important impact of air entrainment within the mix tank was that it resulted in the generation of a hydraulic jump in the slurry discharge pipe immediately downstream of the tank outlet. In effect, this generated a long "standing bubble" along the top of the discharge line and dramatically increased slurry velocity in the bottom half of the 90-cm-diameter steel pipeline. This resulted in rapid thinning of the bottom part of the line (caused by accelerated wear) and, in November 1973, the premature collapse of the discharge pipe after transporting 19 million tonnes of tailing solids. The pipe collapsed because of the low internal slurry pressure brought about by the higher slurry velocity. Tailings flow was diverted into the emergency tailing pond, and the steel discharge line was replaced with a rubber-lined discharge pipe. In addition, steps were taken to reduce air entrainment. Shortly thereafter, detailed modeling design and construction of a new, improved mix tank and outfall line began.

The new mix-tank outfall system is shown schematically in Figure 2.3 and in the photograph in Figure 2.4. This new design prevented air entrainment by taking off much of the static head (6 m) in two on-land drop tanks. As seen in the figures, the slurry was also fed into the tank in a sloping line discharging below the slurry level in the mix tank. In addition, the new mix tank was designed to be a deaerator. By increasing tank diameter to 6 m, residence time was sufficient for any 250-μm-diameter bubble to escape at the open surface against the slowed downward slurry velocity. The mix tank was equipped with a single 107-cm-diameter discharge line with a polyurethane liner.

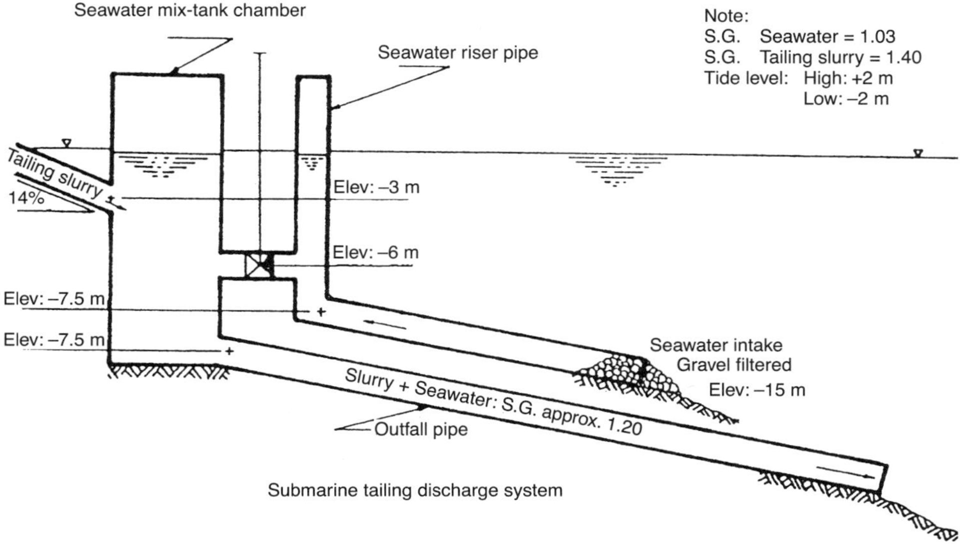

FIGURE 2.3   Schematic diagram of seawater mix tank and outfall system

FIGURE 2.4 Photograph of the seawater mix tank at ICM (operating mix tank is nearest the viewer, with the original kept as a spare)

The new mix tank and outfall line performed very well, having mixed and transported approximately 380 million tonnes of tailing solids up to mine closure. Deaeration was so efficient that no bubble-hood collector was needed. As Figure 2.4 shows, the older system (to right in photo) was maintained for standby needs.

REFERENCES

Aspinall, C. 1995. *The Story of Island Copper*. Madeira Park, BC: Harbour Publishing.

Evans, J.B., D.V. Ellis, and C.A. Pelletier. 1972. The establishment and implementation of a monitoring program for underwater tailing disposal in Rupert Inlet, Vancouver Island, BC. In *Tailing Disposal Today*. San Francisco: Miller Freeman. 512–552.

ICM. 1972. *Summary Report, Pre-Operational Phase, March–September 1971, Environmental Control Program, Rupert Inlet, B.C. ICM*. Vancouver, BC: UBC.

——. 1973. *Summary Report, Initial Production Year, October 1971–September 1972, Environmental Control Program, Island Copper Mine, Rupert Inlet, B.C.* Vancouver, BC: UBC.

Ker, Priestman and Associates, Ltd. 1975. *Study of Tailings Disposal on Land*. Corporate report. Victoria, BC: Ker, Priestman and Associates.

Moore, P.A. 1970. *A Criticism of the Proposed Dumping of Mine Tailings into Rupert Inlet by Utah Mining and Construction Co.* Brief submitted to Director of Pollution Control, Parliament Buildings, Victoria, BC. July 24.

Moore, P.A., C.A. Pelletier, and I. Horne. 2000. Available at www.greenspirit.com (click on "Key Environmental Issues," then on "Waste Disposal: A Case Study in Mining ... Ye Olde Scientific Summary").

North Island Gazette. 1969. Port Hardy, BC. June 18.

Province of British Columbia. 1973. *Pollution Control Objectives for the Mining, Smelting and Related Industries of British Columbia.*

Utah Construction and Mining, Ltd. 1970. *Brief in Support of an Application Under the Pollution Control Act to Discharge Tailings into Rupert Inlet.* Unpublished report.

Waldichuk, M. 1956. *The Disposal of Tailings from Yreka Mine, Neroutsos Inlet.* Internal Report of Fisheries Research Board of Canada, Biological Station at Nanaimo, BC, July 10.

Waldichuk, M., and R.J. Buchanan. 1980. *Significance of Environmental Changes due to Mine Waste Disposal into Rupert Inlet, BC.* Fisheries and Oceans Canada and British Columbia Ministry of Environment.

CHAPTER 3

Engineering Challenges and Solutions

George W. Poling

Construction of the ICM actually began in early 1970, before the public hearing and permitting process was completed. The company gambled that permits would be granted. At the peak of construction, more than 500 people were employed at the site. The project was to cost $70 million, a very large mining project by 1970 standards. Today that same project would cost 5 to 10 times as much.

The orebody was composed primarily of andesitic pyroclastic rocks plus a complexly brecciated and altered swarm of quartz-feldspar porphyry dikes. The principal valuable minerals were chalcopyrite ($CuFeS_2$) and molybdenite (MoS_2). These occurred as finely disseminated crystals, as fracture fillings, and as smears on fractures and slips within the orebody. This orebody was originally estimated to contain 254 million tonnes averaging 0.52% Cu and 0.017% Mo. Gold, silver, and rhenium recovered from this ore also contributed substantially to the economics of the ICM operation. The mineral assemblages of the nonvaluable or gangue components of this ore were quartz (50%–70%), feldspar (2%–20%), biotite and chlorite (5%–10%), magnetite (2%–4%), calcite (~2.5%), pyrite (~3%), sphalerite (0.02%), and galena (0.01%).

Physically, the orebody was located on the northern slope, immediately adjacent to Rupert Inlet as shown in Figure 3.1. Because the optimal method of extracting the ore was with open pit mining technology, the mine is shown as a pit in the figure.

Figure 3.2 shows a composite plan view of the geology of the ICM deposit complete with an outline of the boundary of the final (1995) pit limits (Perello et al. 1995). Figure 3.3 shows a vertical section through the orebody from southwest to northeast as shown on Figure 3.2. Figure 3.3 also shows the ultimate pit limits. These two figures show that the most important geological "structure" in the mine area is the End Creek Fault. These two figures should give the reader a better appreciation of the need to mine perhaps several tonnes of waste rock to get at each tonne of ore. When mining was complete in

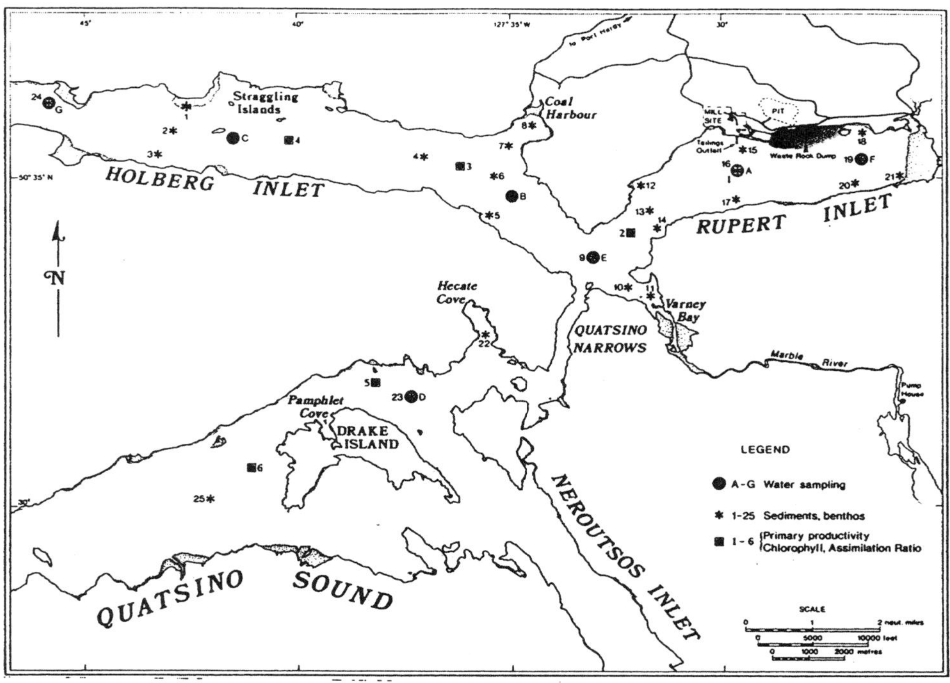

FIGURE 3.1   Map of Rupert–Holberg–Quatsino inlet areas showing location of mine and marine sampling stations

December 1995, the open pit was approximately 2,400 m long × 1,070 m wide and extended to 402 m below sea level—at that time, this was the deepest surface depression on the earth.

MINING

Initially mining was accomplished using five Bucyrus-Erie rotary drills to drill 0.25-m-diameter blast holes nearly 14 m deep. Following blasting, the broken rock/ore was loaded into a fleet of 25 Unit Rig M120 haul trucks of 109-tonne capacity using a fleet of six P&H electric shovels each with 11.5 m^3 buckets. During the first 5 years of operation, 4.1 tonnes of rock (ore and waste rock) had to be mined for each tonne of ore that was delivered to the mineral processing plant or mill. With the mill being fed approximately 30,000 tonnes of ore each day, more than 120,000 tonnes of ore and waste were being mined every day. The waste rock was either stacked in piles on land adjacent to the extremities of the ultimate open pit boundaries or piled along the Rupert Inlet shoreline to form a marine "waste rock dump" as shown in Figure 3.1. Figure 3.4 shows a photograph of an early (double) blast pattern in the ICM open pit mine with on-land waste rock piles in the background.

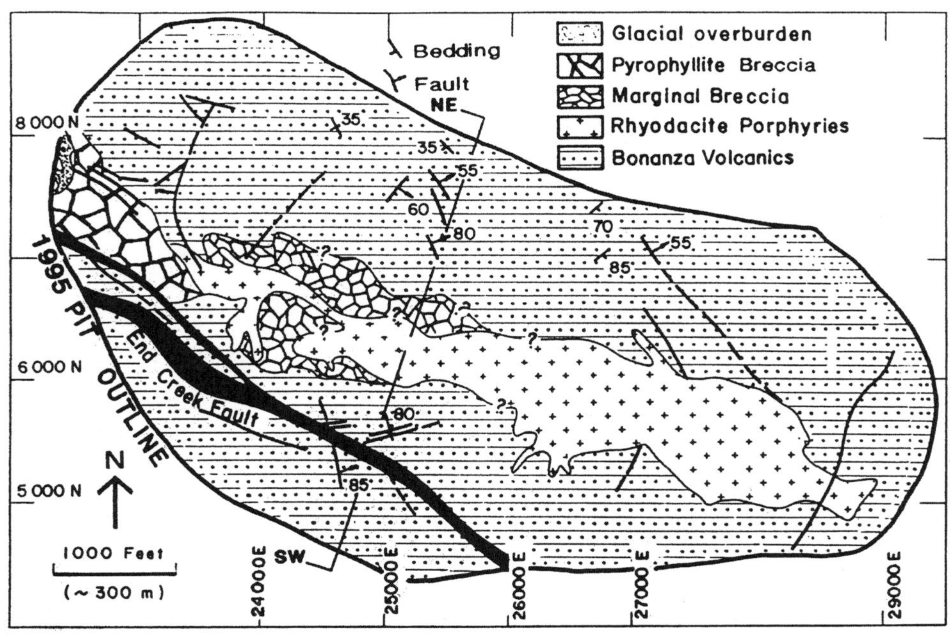

FIGURE 3.2 Geological plan view of open pit showing major rock types

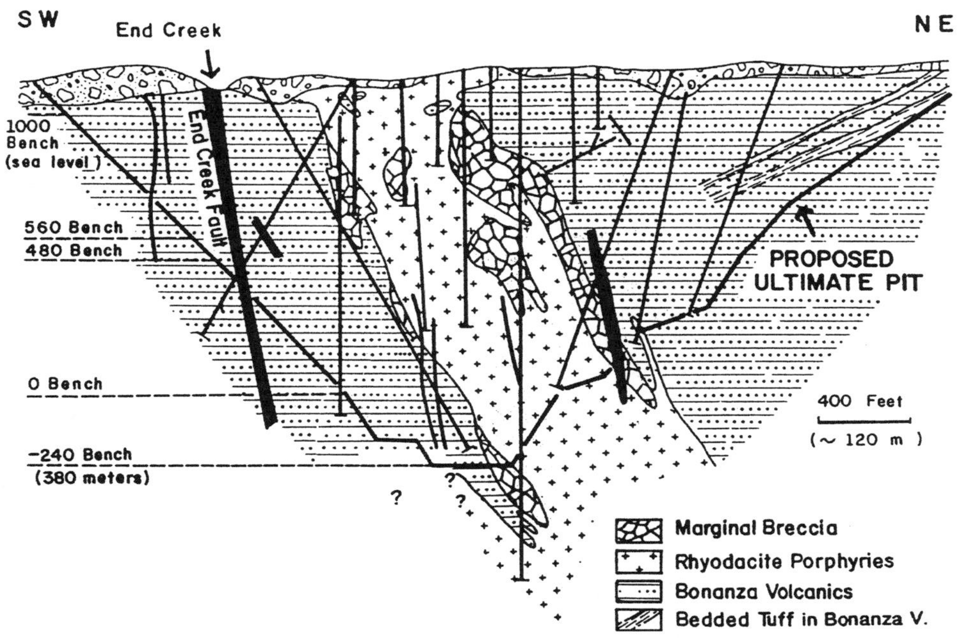

FIGURE 3.3 Geological section through open pit showing major rock types

FIGURE 3.4 Sinking cut blast in open pit mine at Island Copper

MINERAL PROCESSING

Figure 3.5 is a simplified flowsheet for the Island Copper mineral processing plant. This shows mined ore from the open pit, which contains blasted rock from boulder size (maximum 1-m diameter) to fine dust, being delivered by truck to a primary gyratory crusher. This crusher reduced all oversize particles to less than ~0.18 m in diameter. The flowsheet shows that the crushed product was screened (at ~30 cm) and stockpiled in two separate windrows for tunnel reclaim-conveyor feed to six separate large diameter tumbling mills (semiautogenous mills) where the first phase of fine, wet grinding took place. This plant was initially designed for fully autogenous primary grinding (meaning that the large lumps of hard ore would act as the tumbling-grinding media). Approximately 7%–9% by volume of steel grinding balls (~13-cm-diameter) had to be added to the mill charges to achieve design throughput tonnage. This led to the term "semiautogenous grinding" (SAG) for the Island Copper primary grinding mills.

Next, primary ground ore was fed to a series of cyclone "classifiers" (particle size separators), which sorted out particles in the 0- to 100-μm-diameter size range (median grind size = 30-μm [0.03-mm] diameter) for flotation separation. This was the target size range to "liberate" or release the small crystals of valuable chalcopyrite and molybdenite minerals contained within the gangue mineral matrix in the ore. The cyclones in turn returned ore particles coarser than 80 to 700 μm to one of three ball mills (tumbling mills filled with ~40% volume steel grinding balls) for additional or secondary grinding. These rotating SAG mills and ball mills used most of the electrical power consumed in such a plant (typically more than 20 kWh/tonne of ore treated). When operations began, the Island Copper SAG mills were the largest installed in the world. Each mill was

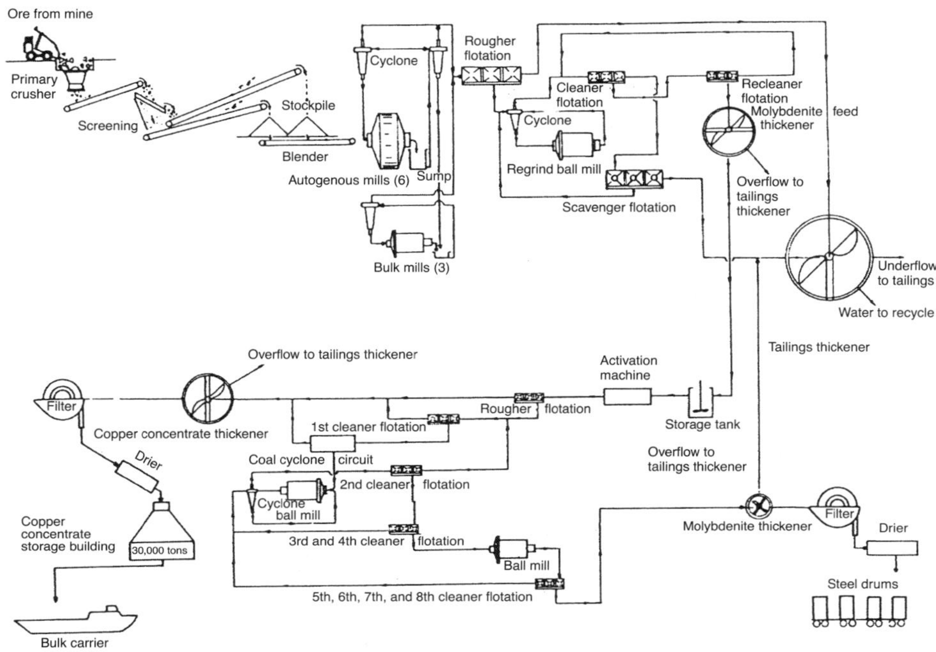

FIGURE 3.5   Simplified schematic flowsheet of Island Copper process plant

installed with 2,235-kW (3,000-hp) motors. Figure 3.6 shows a typical particle size distribution for the finish-ground ICM ore (and tailings). Approximately 50% of this particulate was finer than 30 µm. Knowledge of the size distribution of the tailings and their mineralogy is particularly important in being able to predict the flow and sedimentation behaviors of this material once it is discharged to the marine environment. Further, detailed knowledge of all the processes used to recover the liberated valuable minerals from the finely ground ore is also important to predict potential toxicity of the tailing effluent. In the ICM, froth flotation was the main mineral separation process used.

Froth flotation, in this plant, utilized differences in surface properties of chalcopyrite and molybdenite to separate and concentrate these minerals from the rest of the gangue minerals in the finely ground ore slurry. To achieve this physical separation process, reagents called "collectors" were added to the slurry to adsorb selectively on the chalcopyrite and molybdenite minerals. Adsorption of collector molecules makes their surfaces hydrophobic or air-avid while the rest remain hydrophilic or preferentially wetted by water. By then dispersing fine air bubbles into the stirred slurry in a flotation cell, these bubbles are made to attach preferentially to the hydrophobic particles and buoy them to the surface. So-called "frother" reagents are also added to these flotation cells to ensure that a stable layer of froth or foam covers the surface of the flotation cells. This stable froth layer receives and supports the buoyed-up (floated) particles until they simply overflow a peripheral weir into a froth concentrate launder or trough. Gangue mineral particles (silica, feldspar, pyrite, etc.) that remain hydrophilic stay suspended within the stirred-aqueous slurry and exit as flotation tailings. This flotation process is highly efficient in separating the approximately 1.5% by weight of chalcopyrite and 0.03% of

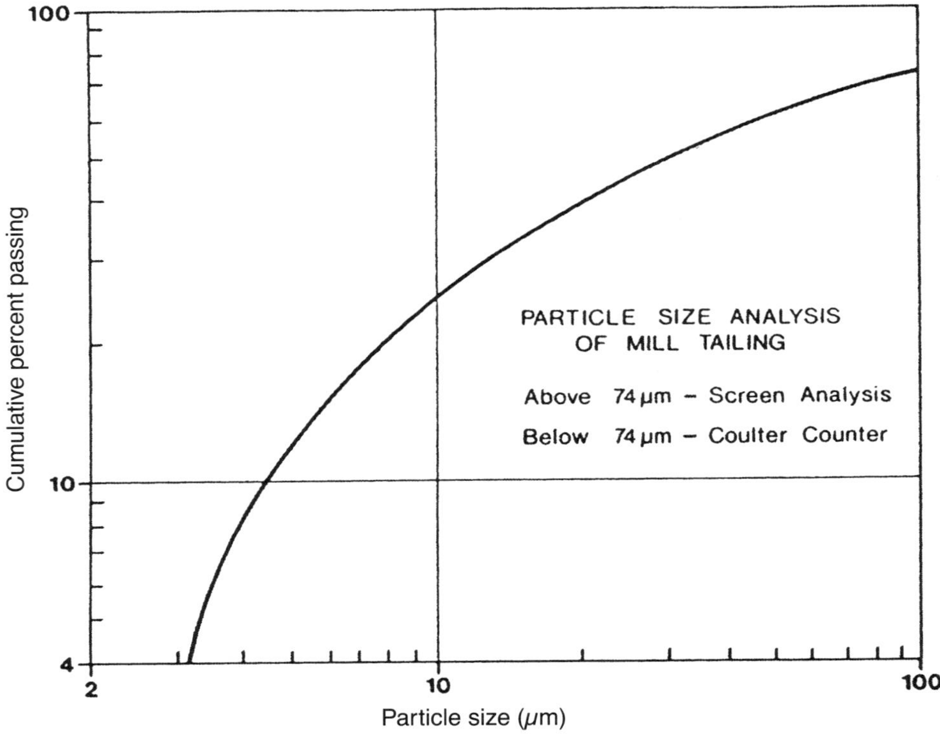

FIGURE 3.6 Particle size distribution of Island Copper ground ore and tailings

molybdenite from the finely ground ore. In this way, approximately 98% of the finely ground ore solids end up as tailings for DSTP.

Copper flotation concentrates that averaged 23%–25% copper metal content (~72% chalcopyrite) were thickened (in a 30-m-diameter thickener), vacuum-filtered, and thermally dried to contain less than 8% water. This concentrate was stored as loose solids in a 32,000-tonne live storage capacity building adjacent to the shipping dock. This concentrate could be loaded directly onto deep sea vessels up to 32,000 deadweight tonnes in displacement via conveyor reclaim at 1,100 tonnes/h. All of the copper concentrates were shipped to smelters in Japan.

Molybdenite concentrate, which typically contained more than 42% molybdenite, was thickened, filtered, and dried before being packaged for sale in 200-L drums. Most of the molybdenite concentrate was sold either in Europe or the United States. The Island Copper molybdenum concentrates contained significant rhenium metal (in substitution for molybdenum in the MoS_2 lattice). Further processing of the molybdenite concentrates brought recovery of this valuable refractory (early in the life of ICM, rhenium was worth >$3,500/kg) and catalytically active metal. Most of the recovered rhenium was actually returned to BHP Billiton so that the company could sell it. The original ore was estimated to contain ½ ppm of rhenium.

Table 3.1 shows typical chemical and mineralogical compositions of the ICM tailings and includes similar analysis of the inorganic components of natural sediments dredged

TABLE 3.1 Typical chemical and mineralogical compositions of tailing solids and natural sediments in Rupert Inlet, British Columbia

| | Contents of Sediments | | | | |
Element or Oxide	A Tailing (%)	B Natural (%)	Ratio A:B	Mineral Species	Tailing Content
SiO_2	62			Quartz	50–70
Al_2O_3	14			Feldspar	2–20
Ca, K, Na, and Mg oxides	10			Biotite and chlorite	5–10
Fe oxides	8			Magnetite	2–4
Fe sulfide	2–3	2–3	1:1	Pyrite	2–4
CO_2	2	—		Calcite	~2.5
Total	98–99	~99	1:1		
	ppm	ppm			
Cu	700	44	16:1	Chalcopyrite	0.2
Mn	650	640	1:1	Mn oxides	n.d.*
Cr	140	125	1:1	In silicates	n.d.
Zn	80	88	1:1	Sphalerite	0.02
Mo	40	2	20:1	Molybdenite	0.01
Co	20	20	1:1	In silicates	n.d.
Ni	20	40	1:2	In silicates	n.d.
Pb	20	25	1:1	Galena	0.002
As	5	5	1:1	Arsenopyrite	n.d.
Cd	3	2	3:2	In silicates	n.d.
Hg	0.03	0.06	1:2	Cinnabar	$>4 \times 10^{-6}$

*n.d.: Not determined.

from the bottom of Rupert Inlet. Close comparison shows these two to be exceedingly similar. (Note also that significant discrepancy exists between the figures for tailings in Figure 3.1 and those provided in Utah Mining and Construction Ltd. [1975]). The tailings contain slightly more residual copper and molybdenum than the natural sediments (they originated from a Cu-Mo orebody). Although similar mineralogically and physically and very similar to natural sediments in terms of particle size, fresh tailing sediments on the inlet floor could be readily distinguished visually. Because they were almost void of organic detritus, fresh tailings were gray in color; natural sediments were deep brown or black.

In the ICM flowsheet, the chalcopyrite and molybdenite were initially floated together to form a "bulk concentrate." Next, the hydrophobicity of the chalcopyrite was surface-chemically destroyed so that the natural hydrophobicity of molybdenite could be employed to subsequently float off molybdenite, leaving chalcopyrite in the slurry phase. Separate copper and molybdenum concentrates, then, were produced for sale at ICM, as indicated in Figure 3.5. This flowsheet shows that other processes such

TABLE 3.2 Typical mill process reagents used at the ICM concentrator

Reagents	Total Consumption (lb)	Mineral Feedstock (tonnes)	Consumption Rate on Treated Material (lb/tonne)	Consumption Rate on Tailings Solids (lb/tonne)
Copper circuit*				
Collector (potassium amyl xanthate)	148,484	20,220,329	0.0073	0.0074
Frother (methyl isobutyl carbinol)	1,730,237	20,220,329	0.085	0.087
Aerodri 100 + 104 (sodium sulfosuccinimate)	11,244	21,340	0.53	0.0006
Polybac E (surfactant)	6,440	1,581,996	0.004	0.0003
Molybdenum circuit				
Arylene M-60 (dioctyl sodium sulfosuccinate)	8,129	12,390	0.66	0.0004
Sodium hydrosulfide	1,622,251	255,028	7.21	0.081
Sodium cyanide	129,361	125,163	1.03	0.0065
Vansene (EDTA)†	65,961	1,399	47.1	0.0033
Gold leach				
Soda ash	25,210	112,792	0.22	0.0013
Hydrogen peroxide	28,155	2,669	10.5	0.0014
Sodium cyanide	45,616	2,511	1,802	0.0023
Activated carbon	2,205	1,689	1.30	0.0001
Production thickener‡				
Alchem 87079 or Superfloc 1202 (flocculent)	30,906	309,186	0.010	0.0015
Tailings thickener§				
Lime (calcium oxide)	23,041,440	20,220,329	1.14	1.15

*Total plant feed: 20,220,329 tons.

†EDTA: ethylenediaminetetraacetic acid.

‡Total copper and molybdenum concentrate produced: 247,479 tons.

§Total tailings: 19,972,850 tons.

as thickening, filtering, and drying were also involved to produce finished dry products for shipping to customers. The above description is certainly oversimplified; many different reagents were employed to achieve efficient physical separation of the mineral products at ICM. Most of these "milling chemicals" were added in minute amounts, and most would also report mainly to the mineral concentrates. Table 3.2 presents a far more complete listing of typical milling chemicals and consumption rates used at Island Copper. Because some portion of each of these reagents undoubtedly reported to the tailings, care was taken to ensure that the tailings effluent was nontoxic. In ICM's 25 years of operation, only one of the hundreds of monthly tailing samples exhibited toxicity in their routine 96-hour bioassay tests at 100% concentration. This resulted from mill testwork using a new reagent that itself had not been separately bioassayed. No toxicity was found in the receiving waters, undoubtedly because of the high inherent seawater dilution of the liquid phase of the tailing. After the one unfortunate incident with the new reagent, more care was taken to make sure that no new reagents were mill tested without prior toxicity testwork. This occurrence does, however, point out the need for continuous vigilance when utilizing DSTP.

CAPITAL AND OPERATING COSTS

Although the ICM and plant was originally estimated to cost CDN $70 million, the actual capital outlay by late 1971 was CDN $88 million (Aspinall 1995). The first copper concentrate ship left the Island Copper dock in December 1971, only 30 months after Utah Mining announced its decision to proceed with construction. In December of 1971, this mine employed 120 salaried staff members plus 520 people who were paid by the hour.

Although the two large-diameter (114-m-diameter) tailing thickeners, shown in Figure 3.5, might appropriately be classified as part of the tailing placement system, their main objective was to reclaim approximately 50% of the process water requirements for recycle to the plant. In this instance, then, including the tailing thickeners as part of the DSTP system's capital cost is probably not justified. The DSTP system comprised the 0.86-m-diameter (initially) steel (and later high-density polyethylene) pipeline. The pipeline, approximately 900 m long, carried thickened tailings to the shoreline seawater mix tank and then along the sloping seabed via a pipeline to discharge at −50 m. The completed DSTP system is estimated to have cost CDN $1 million.

Additional initial capital costs included a well-equipped environmental laboratory complete with a clean room. This enabled accurate analysis to be conducted on-site at the parts-per-billion level demanded for the environmental monitoring. Additional capital cost of the environmental laboratory was on the order of CDN $600,000 (in 1970). In 1974 Utah Mining was asked to prepare an engineering report on an alternative scheme for an on-land tailing disposal facility. The capital cost for such a facility, complete with dual pumping stations (because of the high static lift of 110 m), was CDN $24.2 million (Utah Construction and Mining, Ltd. 1975). Operating costs of such an on-land alternative would also have been much higher than the DSTP system because of the high cost of pumping an abrasive slurry up over a 110-m lift and over several kilometers of pipe length. In comparison, the tailings flowed downslope via gravity to the DSTP outfall. The most significant operating cost of the DSTP system was probably the oceanographic monitoring, evaluation, and reporting costs, which amounted to less than CDN $1 million/year.

REFERENCES

Aspinall, C. 1995. *The Story of Island Copper*. Madeira Park, BC: Harbour Publishing.

Perello, J.A., J.A. Fleming, K.P. O'Kane, P.D. Burt, G.A. Clarke, M.D. Himes, and A.T. Reeves. 1995. *Porphyry Copper-Gold-Molybdenum Deposits in the Island Copper Cluster, Northern Vancouver Island, BC*. Canadian Institute of Mining, Metallurgy, and Petroleum (CIM) Special Volume 46. 214–238.

Utah Construction and Mining, Ltd. 1975. *Study of Tailings Disposal on Land*. Vancouver, BC: Study conducted by Ker, Priestman and Assoc. Ltd.

CHAPTER 4

The History of the Morphological Change on the Seafloor of Rupert and Holberg Inlets

James W. Murray and Clem A. Pelletier

BACKGROUND

The ICM, an open pit copper-molybdenum mine, began operating in October 1971 on the north shore of Rupert Inlet. Between its start-up in 1971 and the end of December 1995, some 358 million tonnes of tailings were discharged into Rupert Inlet through an underwater tailings placement (UTP) system that discharged at 50-m depth. During the mine's operational period, mined rock was also placed along the north shore, creating about 262 ha of new land from Rupert Inlet. Waste rock and underwater placement of tailings have also changed the morphology of the inlet. The mine was closed in 1995 after exhausting the orebody.

Located toward the northern end of Vancouver Island, Rupert Inlet is a fjord 10 km in length, connected to the Pacific Ocean through Quatsino Narrows and Quatsino Sound (Figure 4.1). The inlet is adjoined by 34-km-long Holberg Inlet to the west. The circulation of Rupert Inlet is driven by tides, freshwater flow from the drainage basin, and the 18-m-deep sill of Quatsino Narrows. The tide is mixed semidiurnal with tidal ranges varying from 1 to 4 m. The inlet is well mixed (Drinkwater and Osborn 1975) in contrast to other inlets in British Columbia (Pickard 1963).

From 1969 to 2000, ICM conducted a series of preoperational, operational, and post-operational bathymetric and continuous seismic profiling surveys, as shown in Table 4.1.

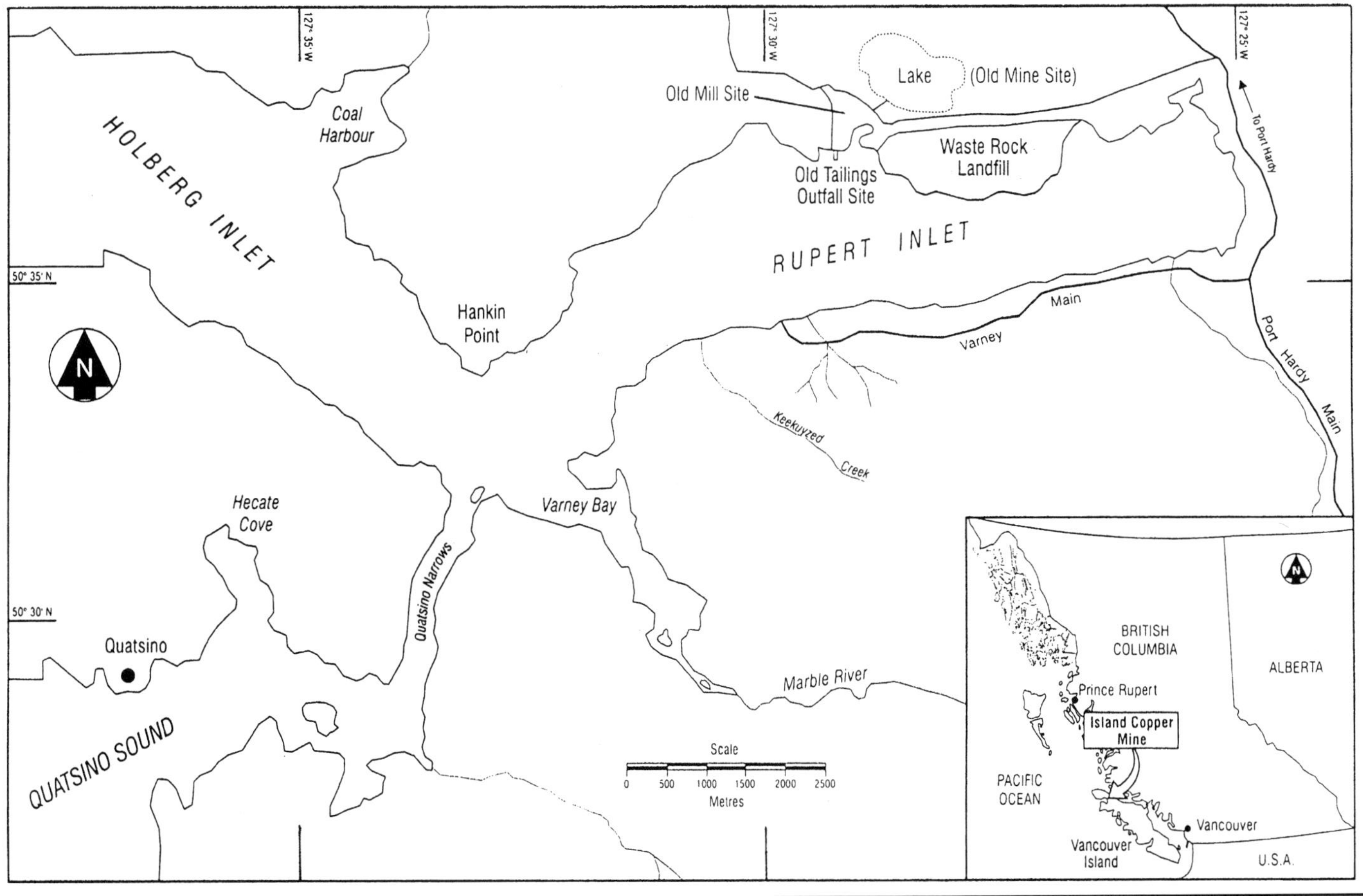

FIGURE 4.1 Rupert Inlet and ICM site

TABLE 4.1 Bathymetric and continuous seismic profiling surveys of Rupert Inlet from 1969 to 2000

Bathymetric Surveys	Continuous Profiling Surveys
McElhanney Survey & Engineering Ltd.	
February 1969	
Canadian Hydrographic Survey (CHS)	
June 1971	
UBC*	
November 1976	**UBC**
September 1977	March 1971
December 1977	November 1972
September 1978	November 1973
February 1979	November 1974
August 1979	October 1975
CHS	January 1977*
October 1982	October 1988
ICM	**ICM**
January 2000	January 2000

*Only covered part of area.

PURPOSE

This chapter presents the results of these investigations into the development, over a 29-year period, of the marine geomorphological changes that occurred in Rupert and Holberg inlets. These studies have shown that the discharge of mine tailings into Rupert Inlet led to the formation of a submarine channel system that transported and deposited tailings from the underwater discharge point down the length of Rupert Inlet into the Hankin Point area (Figure 4.1) and beyond into Holberg Inlet. This study is one of the few instances in which evolution of a submarine channel system has been monitored over such a long period of time.

Carstens and Tesaker (1972), Tesaker (1975), and Normark and Dickson (1976a, 1976b) conducted previous studies of subaqueous channel development resulting from mine tailings deposits. The Rupert/Holberg depositional system offers an opportunity to study turbidity currents and channel development in real time, at scale orders of magnitude greater than achievable in the laboratory, and under conditions in which the source of sediment is quasi-steady stable in terms of rate, type, and size of material discharged. Superimposed on this quasi-steady input, however, is the irregular occurrence of surge-type turbidity currents, which result from the slumping of previously deposited tailings. The relative importance of continuous and surge-type turbidity flow can therefore be regarded as an important aspect of tailings transport and dispersal in these systems. Current measurements (Johnson 1974) and acoustic remote sensing studies (Hay, Burling, and Murray 1982; Hay 1982, 1983) of both surge-type and continuous flow turbidity currents in Rupert Inlet are well documented. The acoustic remote sensing data clearly indicated that channel overspill occurred in both types of flow. Previously, Holtedahl (1965) reported on naturally formed turbidities in Hardanger Fjord in Norway, which set the foundation for this research.

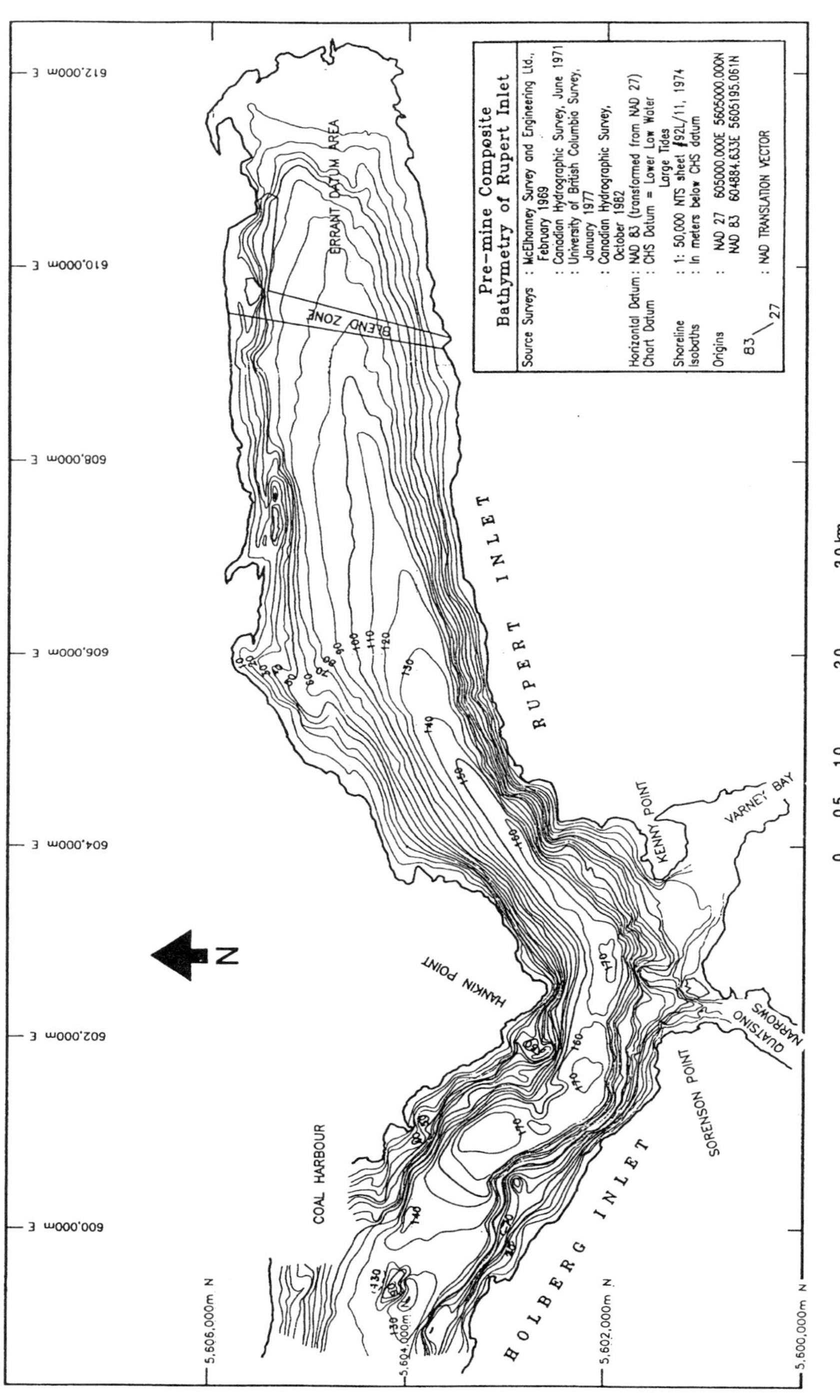

FIGURE 4.2 Pre-mine composite bathymetry of Rupert Inlet

Approach

The bathymetric evolution and morphological changes were determined by comparing the pre-mine (Figure 4.2) and the post-mine time-series bathymetric maps (Figure 4.3) and by utilizing the seismic profiles to assess the amount of material deposited on the seafloor during the life of the mine. This has been done qualitatively and quantitatively by:

- Superimposition of isobaths
- Superimposition of nine bathymetric cross-sectional profiles
- Creation of an isopach map of sedimentation
- Estimation of beach dump volume from the isopach map
- Superimposition of hypsometric curves of the inlet (hypsometry refers to a graphical measure of volume, or cross-sectional area, as a function of depth)
- Superimposition of nine hypsometric curves of cross-sectional areas

The total volume of deposition is estimated, and its distribution is shown by an isopach map and hypsometric relations. Isopach maps are prepared using both the bathymetric and the seismic surveys. A discussion is presented on the nature of the changes in the isobaths.

FIELD INSTRUMENTATION AND SURVEYS

Bathymetric Surveys

Table 4.2 summarizes various survey parameters and presents facts about the pre-mine survey, the surveys conducted during mine operations, and the post-closure survey of January 2000 (Rescan 2000).

A DGPS was used for the bathymetric survey (Rescan 2000). The Global Positioning System (GPS) is a satellite-based system that uses the WGS-84 as a positioning datum. WGS-84 was internationally adopted as the reference frame for broadcast orbits on January 29, 1997. The real-time differential correction for the survey was ensured by the Coast Guard DGPS beacon located at Alert Bay. Positions are accurate to 1 m root mean square (rms).

The soundings for January 2000 were made using an Odom Echotrak 200-kHz echo sounder mounted aboard the vessel *High Rye*. North-south (cross-section) sounding tracks were made at about 100-m intervals (Figure 4.4). Sounding and GPS data were merged into the GPS data logger using Asset Surveyor software to position each sounding track accurately. The accuracy for a narrow beam echosounder is ±0.01% of the sounded depth.

Some cross-sectional profiles were taken across the inlet to illustrate the sounding and seismic profiles and to assess the morphological change (Figure 4.5). An example of a sounding profile through cross section 6 is shown in Figure 4.6.

Seismic Surveys

The continuous seismic profiling surveys were conducted using either an EG&G Boomer or Uniboom Seismic Profiler. Sounding signals were generally fired at 200 J at every 300 ms and a hydrophone streamer received the reflected signals. Signal processing was carried out by a Krohnhite filter set with a pass band of 0.6 to 15 kHz with a gain of 20 db.

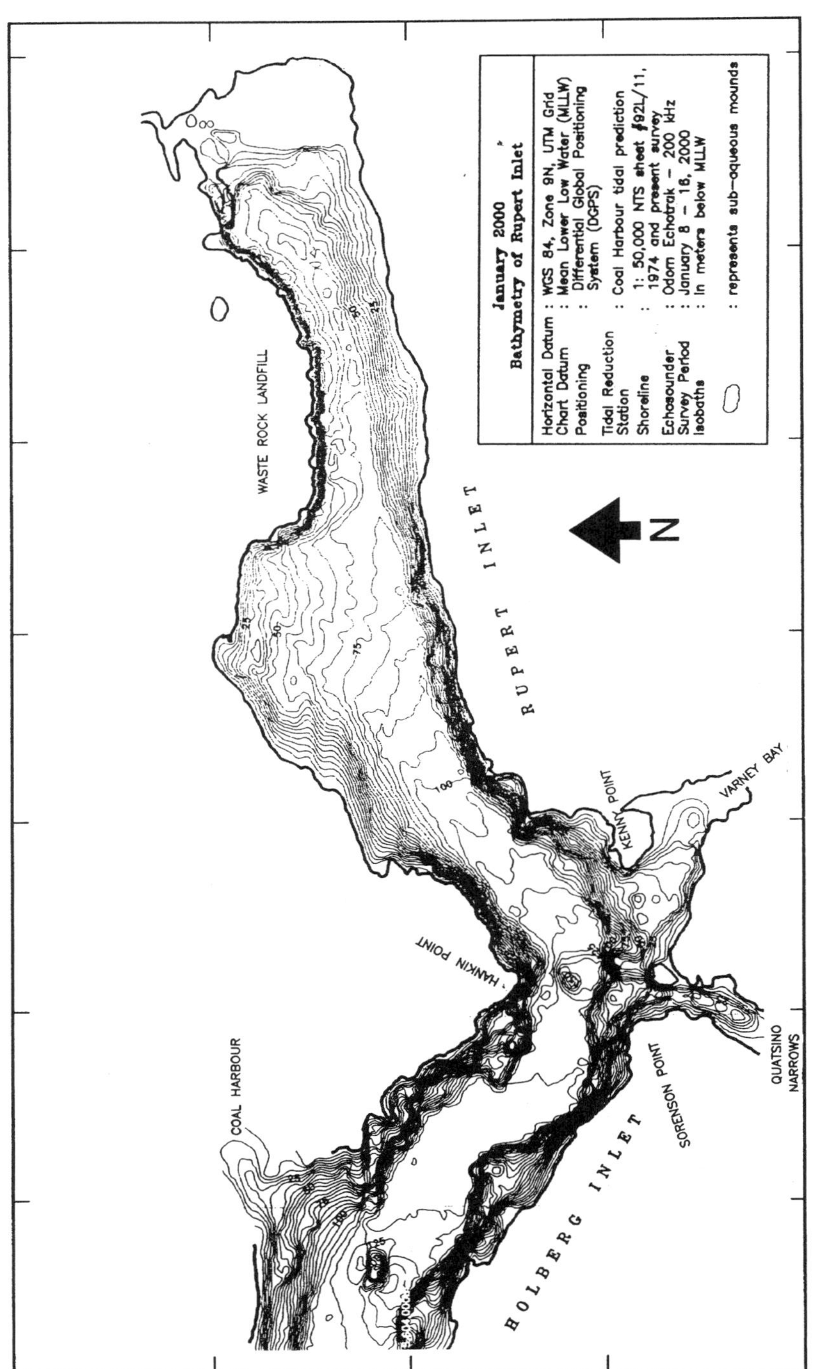

FIGURE 4.3   Bathymetry of Rupert Inlet (January 2000)

TABLE 4.2 Summary of survey parameters

Parameter	Pre-mine Composite Map	Rescan 2000 Survey
Time	Composite bathymetry compiled from 1969, 1971, 1977, and 1982 surveys	January 2000
Horizontal datum	NAD 27 Grid[*]	WGS-84, Zone 9N, UTM Grid
Positioning	Theodolite/Sextant Mini-Fix Range-Range System	DGPS
Vertical datum	CHS = LNT = LLWLT	MLLW
Tidal reduction station	Coal Harbour Prediction	Coal Harbour Prediction
Reference shoreline	1:50,000 NTS sheet # 92L/11, 1974	1:50,000 NTS sheet # 92L/11, 1974
Echosounder	Various	Odom Echotrak—200 kHz

[*]The old North American Datum 27 model (NAD 27) was transferred to the more recent NAD 83 model, and all the survey coordinates are expressed in meters relative to a datum.

Acronyms:

DGPS: Differential Global Positioning System

LLWLT: lower low water large tides

LNT: lowest normal tides

MLLW: mean lower low water

NTS: National Topographic System

UTM: Universal Transverse Mercator

WGS-84: World Geodetic System of 1984

The surveys utilized an EPC 9800 line scan recorder set on a 150-ms sweep. Figure 4.7 shows the seismic profile lines across Rupert/Holberg Inlet as of January 2000.

The seismic equipment was supplied and operated initially by UBC and subsequently by Terra Remote Sensing Inc. of Sidney, British Columbia. Profiling was done along several track lines, both across and along the inlet, by maintaining a vessel speed of 3.5 to 4.0 knots, varying often because of tidal current. As described earlier, positioning was done by GPS.

Earlier seismic surveys established by ICM staff and conducted in 1971, 1972, 1973, 1974, and 1975, used dead reckoning and shoreline markers. Subsequent seismic surveys utilized the Mini-Fix Range-Range System, which provided seismic survey positioning.

All the seismic profiles obtained over the years show depositional thickness and its spatial distribution both across and along Rupert/Holberg Inlet. In addition, the depositional and transport patterns are noticeable in all profiles. Figure 4.8 shows the seismic profile along line 19, taken during the January 2000 survey. Figure 4.9 is a line drawing interpretation.

DATA PROCESSING AND MAP PREPARATION

To process all the bathymetric and seismic survey data, three tasks were necessary:

- Preparation of the bathymetric map based on the January 2000 survey
- Transformation of the pre-mine composite bathymetry to the present bathymetric conditions
- Processing of the current seismic survey to supplement previously obtained data

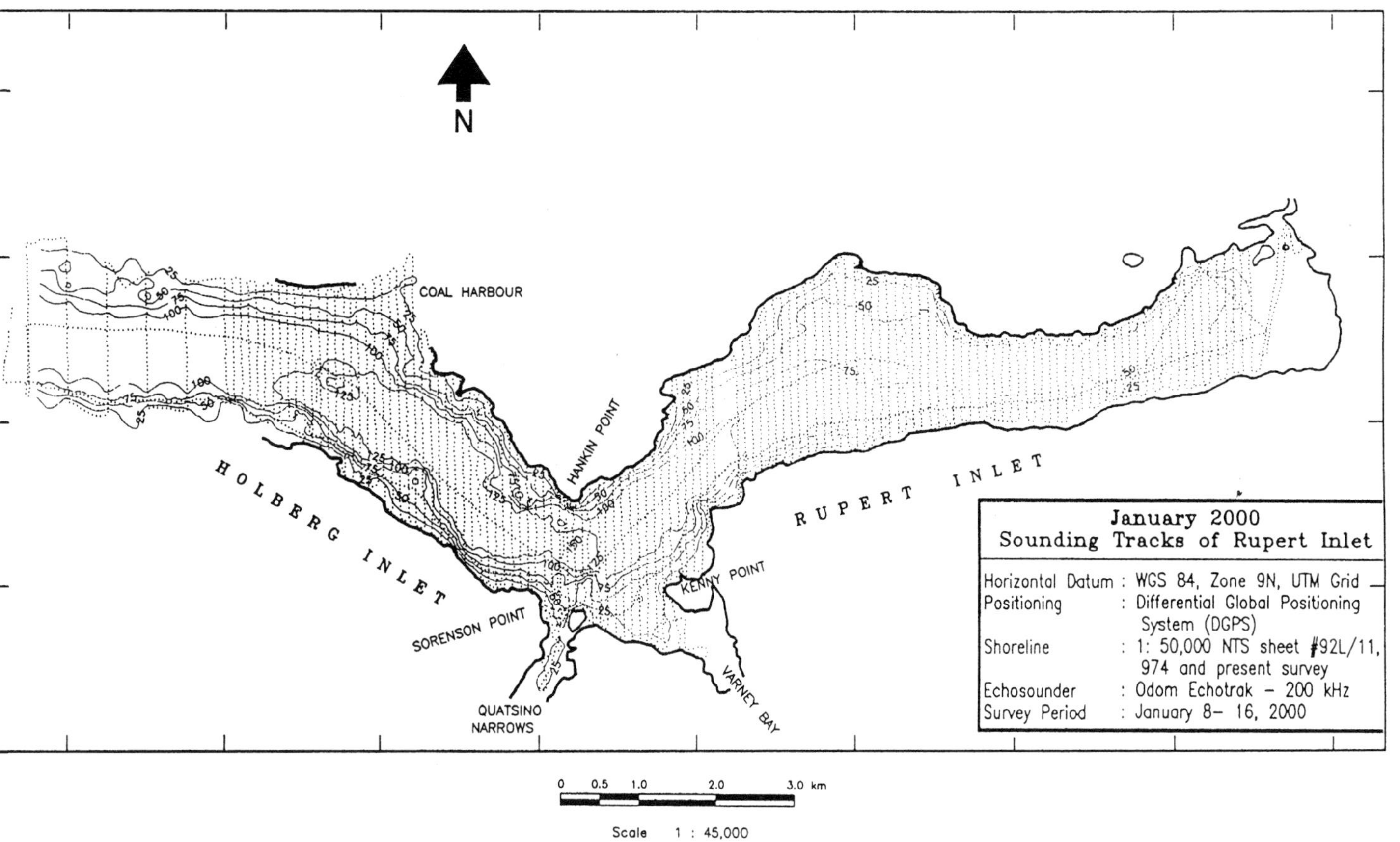

FIGURE 4.4   January 2000 sounding tracks of Rupert Inlet

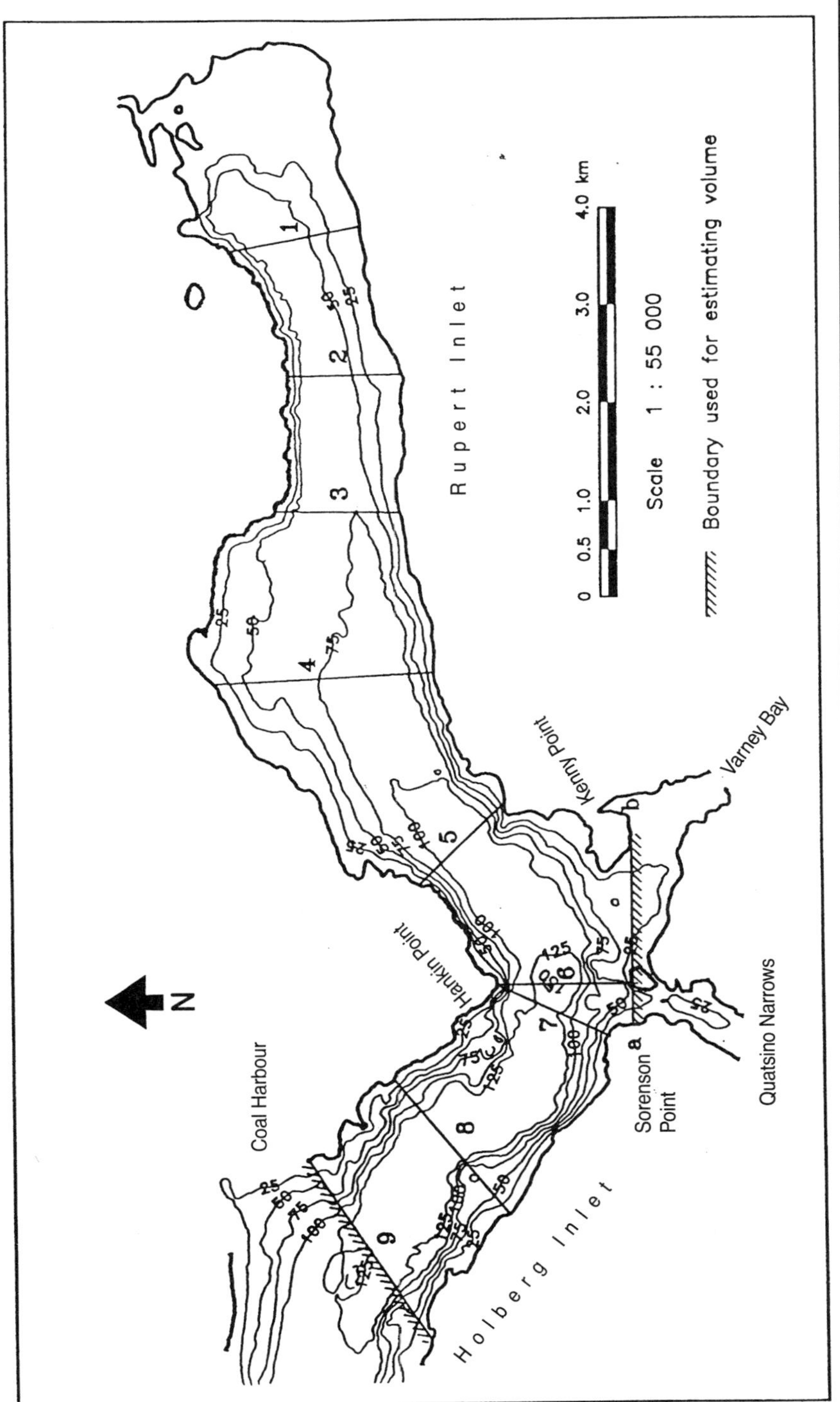

FIGURE 4.5   Locations of cross sections

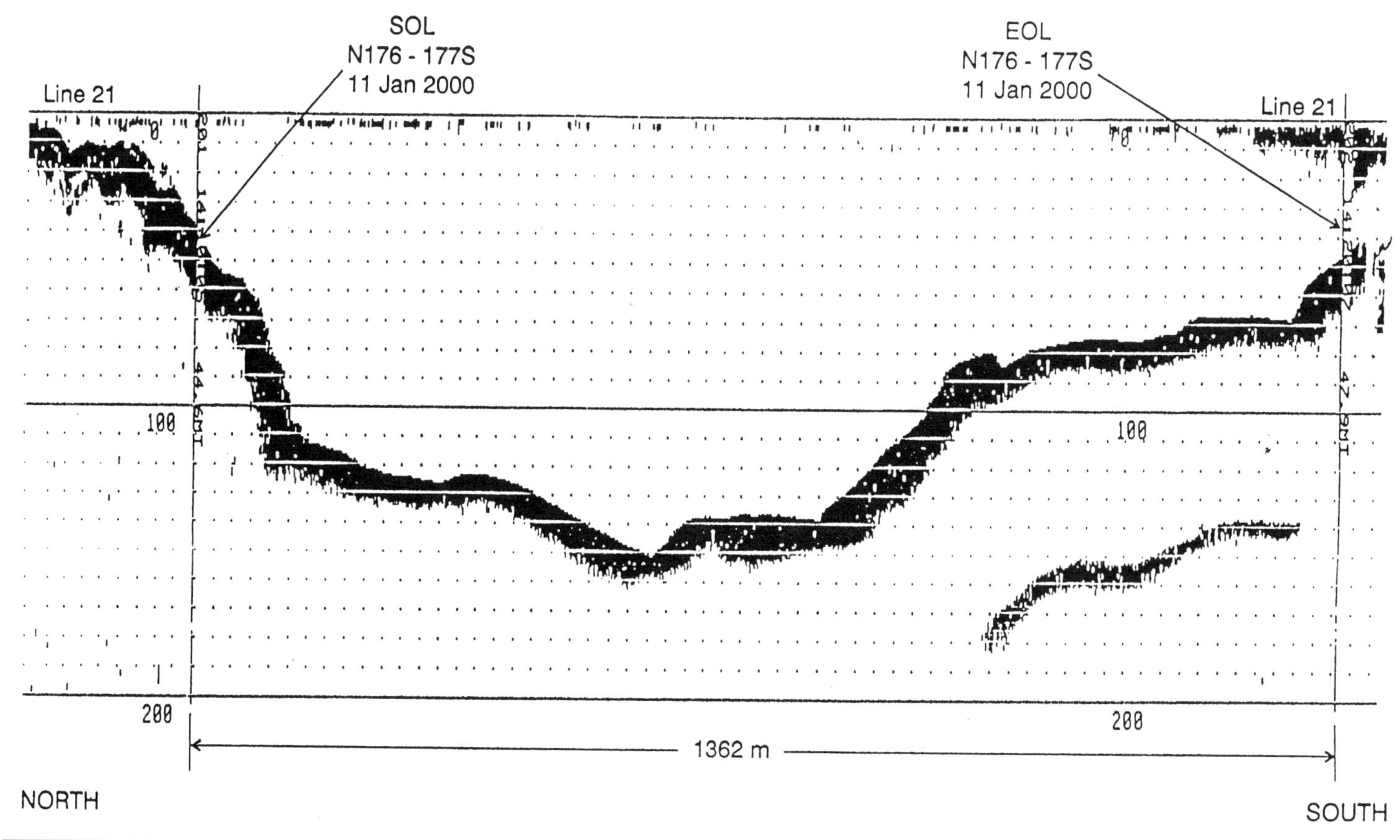

FIGURE 4.6   Sample sounding profile along cross section 6

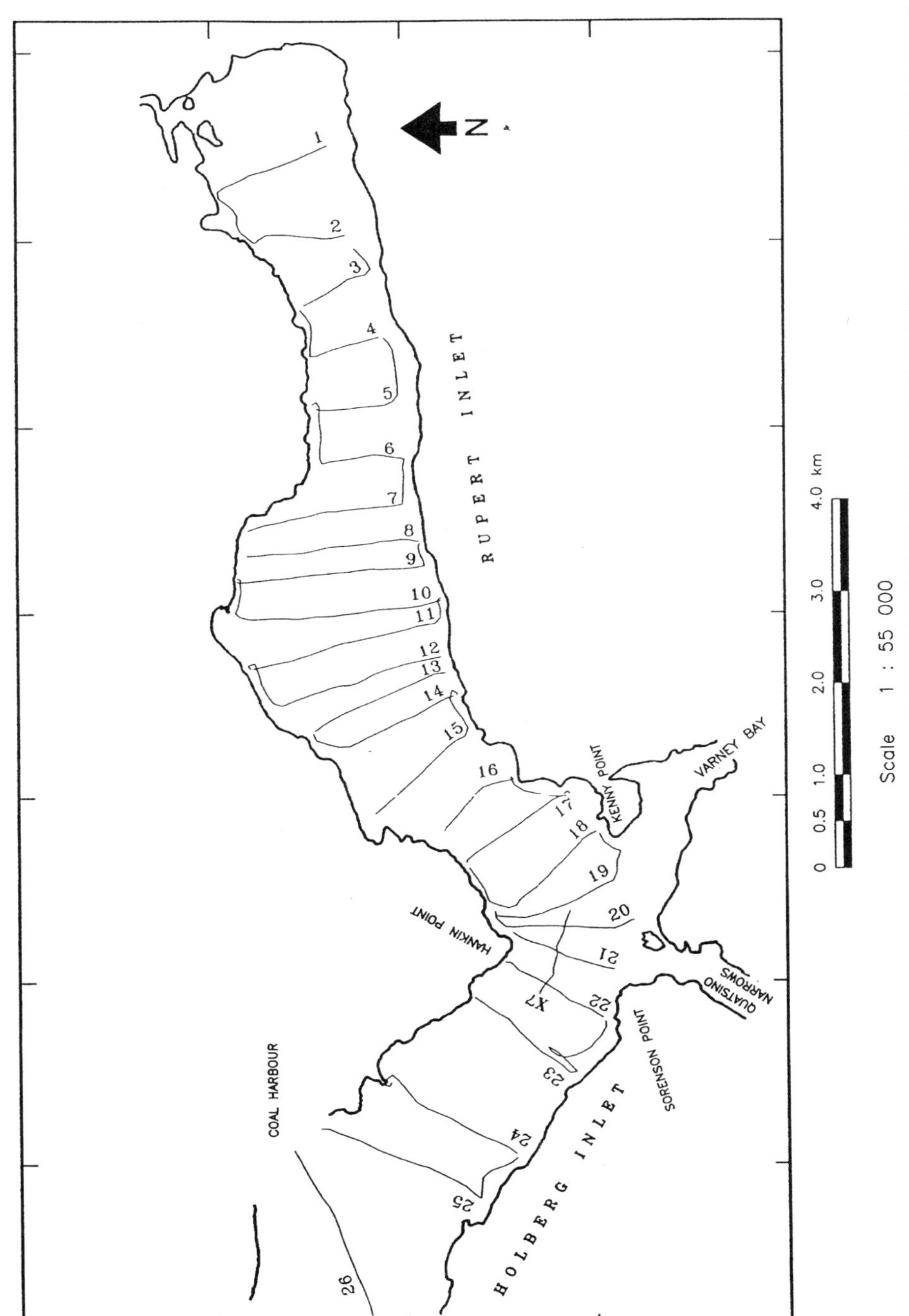

FIGURE 4.7   Seismic lines across Rupert/Holberg Inlet

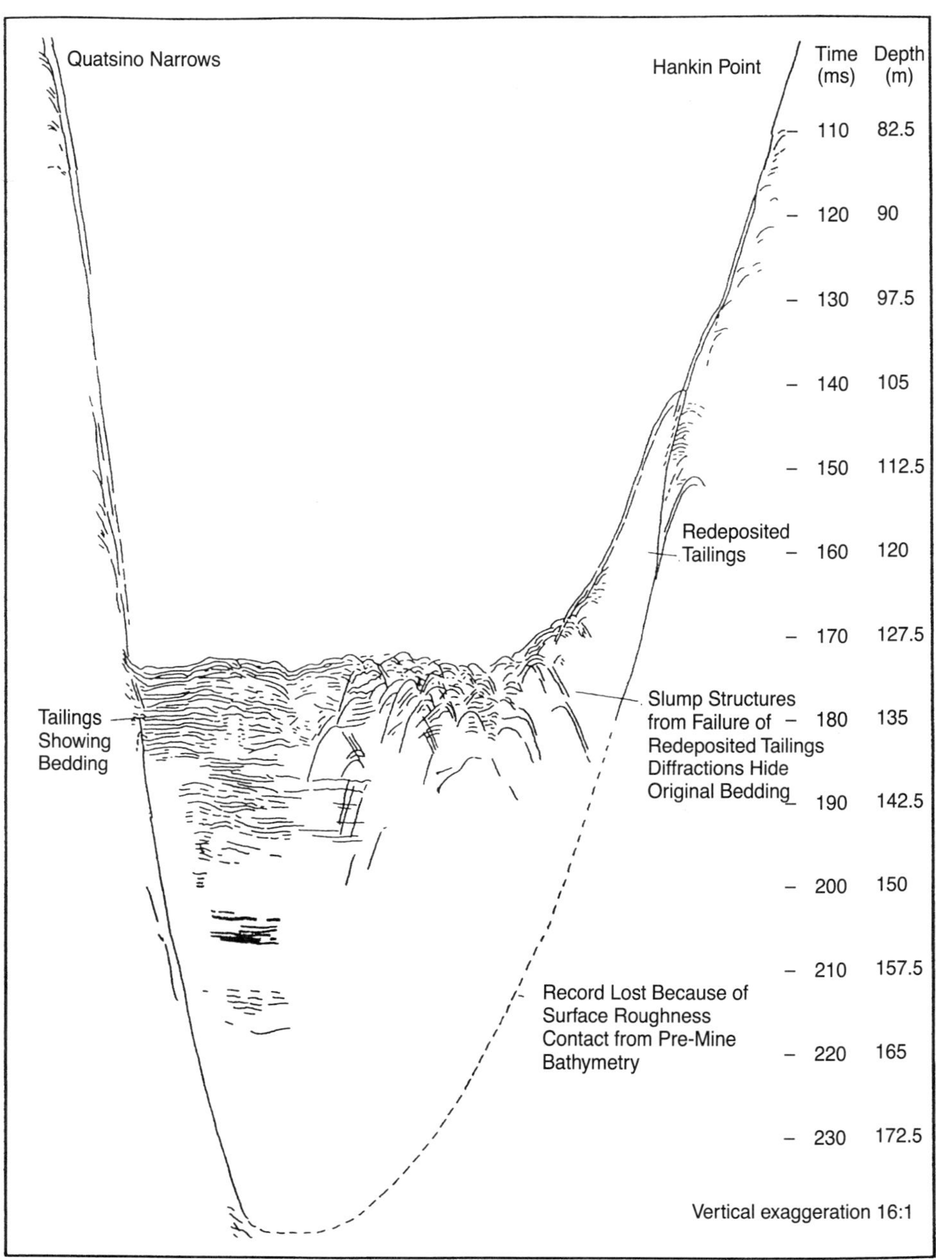

FIGURE 4.8   Seismic profile through line 19

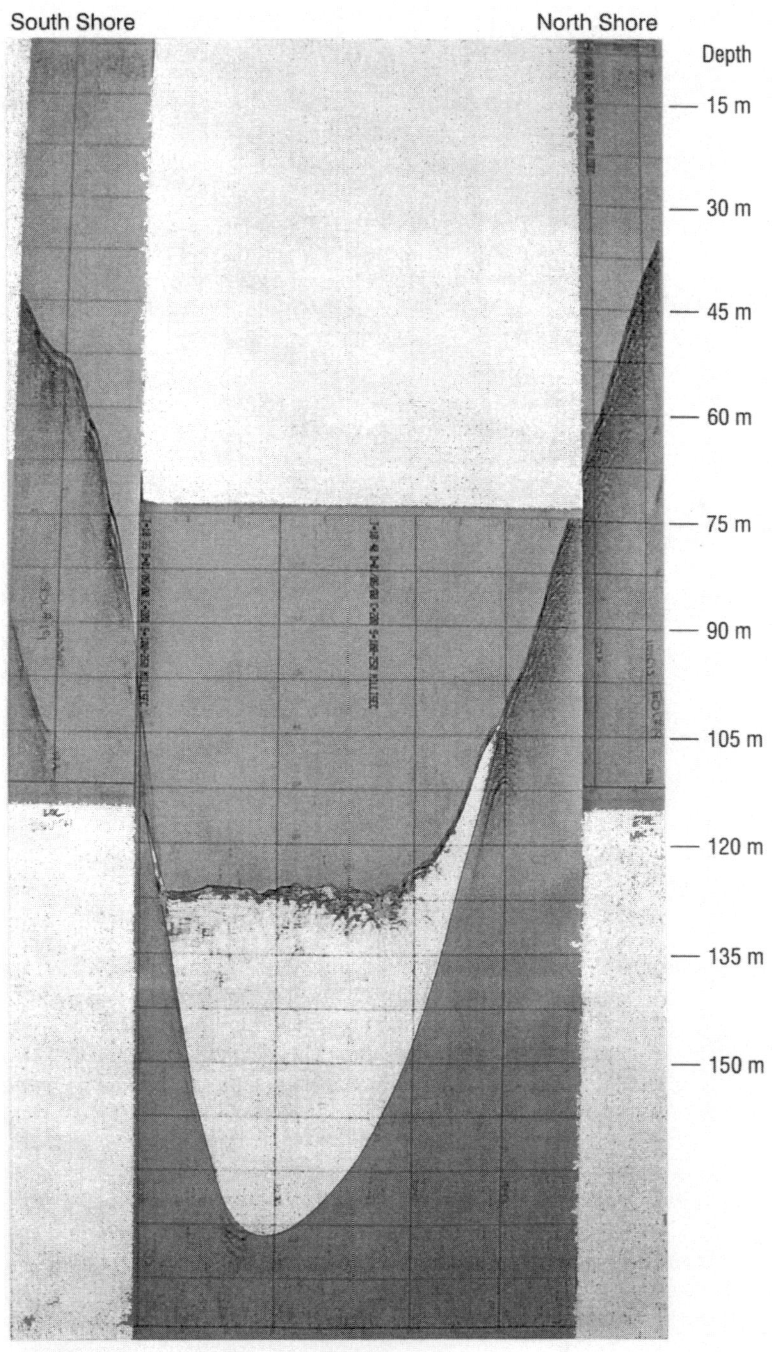

FIGURE 4.9 Rupert Inlet seismic profile number 19

For the January 2000 survey, the sounded depths were first checked for spurious electrical noise, and any such abnormalities were removed. A tidal correction was then applied to reduce the validated depths to chart data. The selected chart datum is MLLW. The predicted tidal height at Coal Harbour (127°35.0'W; 50°36.0'N) (refer to Figure 4.1) was used for this purpose. Coal Harbour is a secondary tidal station—the predictions at this station are based on corrections applied to Winter Harbour (128°2.0'W; 50°31.0'N) tides. The higher high water and the lower low water at Coal Harbour lag by 58 minutes and 57 minutes from those of Winter Harbour, respectively. The height corrections are −9 cm and +9 cm for higher high water and lower low water, respectively. According to the CHS tide tables, the mean level at Coal Harbour is 2.2 m above LLWLT. Defined as the average of the lowest waters (one from each year of the 19-year averaging period), LLWLT is equivalent to LNT. As mentioned earlier, the January 2000 survey is referenced to WGS-84 horizontal datum.

It is assumed that the horizontal datum of the pre-mine bathymetric survey was referenced to NAD 27. In order to make comparison with the present survey, the NAD 27 horizontal coordinates were transformed into NAD 83 coordinates using a standard transformation software package. As Figure 4.2 shows, a small area at the head of Rupert Inlet did not belong to any known horizontal datum, so related observations were made. For the scale of this survey, the NAD 83 and the WGS-84 are nearly identical. Therefore, a consistency in horizontal datum was reached between the January 2000 survey and the earlier composite survey.

In absence of any documented information, the chart datums for all previous surveys are assumed to be the CHS datum that is defined as LNT (Thomson 1981). According to CHS (1999), the CHS datum represents LLWLT. The present survey datum is MLLW, which is the accepted chart datum along the U.S. Pacific West Coast. Based on CHS definitions, the MLLW is slightly higher than the Canadian CHS datum (Thomson 1981). For all practical purposes, they are treated as identical.

RESULTS AND DISCUSSION

Bathymetry of the Inlet Systems

The pre-mine composite bathymetry of Rupert Inlet is shown in Figure 4.2 in meters below CHS datum. It shows that Rupert Inlet is a rather asymmetric trough with the north inlet floor sloping southward. The axis of the channel is located approximately two-thirds of the way across the inlet. The channel axis slopes smoothly down-inlet to reach the deepest point of 170 m off Hankin Point near the entrance to Quatsino Narrows. From the Hankin Point area, the floor of adjacent Holberg Inlet then slopes gradually upward to the northwest shoreline 130 m off the entrance to Coal Harbour. Essentially, it was an asymmetric U-shaped trough sloping southwesterly to the deepest point at Hankin Point. Into this setting, ICM began depositing tailings in October 1971.

In contrast, Figure 4.3 shows the results of a January 2000 bathymetric survey taken about 4 years after the mine closed in December 1995. The isobaths representing depths below MLLW are plotted at 5-m intervals. The principal changes to note are the displacement of the north shore of Rupert Inlet to the south by the waste-rock landfill. In addition, the inlet shoaled considerably. For example, 2 km directly south of the mine site, the water shoaled from 130 m to 83 m with a net result of producing a more flat-bottomed inlet compared to the original, asymmetric, U-shaped profile. Also note that the deepest point in the hole off Hankin Point was 20 m shallower by January 2000, but this will be discussed more in a later section. Overall, there is a southwestern shift of the isobath resulting from the transport and deposition of tailings down-slope and from the shoaling produced by the waste-rock landfill. In Figure 4.9, note that there are also at least 2 subsidiary channels pointing downslope to Hankin Point as originally observed by Hay (1982).

Original Prediction of Tailings Placement in the Rupert/Holberg Inlet System

Once a decision was made to consider placement of tailings from the ICM into the Rupert/Holberg Inlet System, the original proponent of the mine, Utah Mining, retained a project team to predict where the tailings would end up. In a brief presented to the British Columbia Pollution Control Board Hearing (January 1971), the team predicted that the tailings would spread principally downslope in Rupert Inlet covering much of the deeper floor of Rupert Inlet below 30 m, down-inlet to Hankin Point, and beyond into Holberg Inlet. This prediction turned out, in retrospect, to be remarkably accurate, with one exception. The prediction did not foresee that the strong upwelling currents at Hankin Point would lift some of the tailings off the bottom up into the euphotic zone and in fact be visible at the surface as milky-white, turbulent boils. Originally, Pickard (1963), followed by Drinkwater and Osborne (1975), Stucchi and Farmer (1976), and Stucchi (1980), studied the physical oceanography of the system. This upwelling and deep water renewal is controlled primarily by the buoyancy of the incoming jet of ocean water entering through the narrow, shallow sill (30 m) of Quatsino Sound and the run-off from the Marble River (Stucchi 1980).

Johnson (1974) studied the distribution of original inlet sediments and the subsequent discharge of tailings on top of the original sediments. He concluded that during deep-water renewal, sediment/tailings were eroded from the deepest part of the basin near Hankin Point and redistributed in the inlet system. Johnson also made near-bottom current measurements subsequent to the start of tailings discharge and observed several short-lived events of high-amplitude, down-inlet currents (up to 50 cm/sec), which may have been turbidity currents.

Continuous Seismic Profiling Surveys

Continuous seismic profiling (CSP), as previously discussed, was conducted in 1971, 1972, 1973, 1974, 1975, 1977, 1988, and 2000. These surveys added a third dimension to the previously discussed bathymetric surveys. The first CSP survey in 1971 showed the shape and thickness of postglacial sediments on the floor of Rupert Inlet and yielded information from which to infer depositional processes within the inlet.

Pre-mine Deposition Patterns in Rupert Inlet

The March 1971 CSP surveys showed that sediment was asymmetrically banked part way up the northwest wall of Rupert Inlet, forming a wedge-shaped deposit that thinned to zero adjacent the south side of the inlet (Johnson 1974). These sediments thinned dramatically down-inlet, reaching a minimum thickness in the Hankin Point area before starting to thicken again in Holberg Inlet. Further, Johnson used information from a bottom-sampling program to show a dramatic increase in sediment grain size down-inlet to a maximum of pebbles and cobbles off Hankin Point.

Once mining started and tailings were being deposited into Rupert Inlet, the CSP surveys became a very useful tool for differentiating mine tailings from the previously deposited natural sediment.

Deposition of Mine Tailings

ICM began to discharge tailings in October 1971. Tailings were, in the first 10 years, discharged at a rate of about 380 kg s^{-1} from a pipe 1.07 m in diameter as a slurry of solids, freshwater, and seawater in the ratio 1:4:5 parts by volume. The volume discharge rate was 1.4 m^3 s^{-1}. This volume of seawater was drawn from a depth of 15 m, and the discharged point was 49 m below the surface and 1 m above bottom (see outfall, Figure 4.1). The particles constituting the tailings have a median diameter of 0.003 mm. Approximately 65%–75% of the tailing solid particles are smaller than 0.074 mm (Poling 1982). Waste rock was dumped into the inlet at the waste-rock landfill (see Figure 4.1) at a rate approximately three to four times that of the tailings solids (ICM Annual Assessment Reports covering the period from 1971 to 1999, inclusive). These reports represent the single most extensive time-series reports ever undertaken on any fjord.

The 1972 CSP survey showed that, initially, tailings were deposited as conical mounds off the end of the outfall. Once the tailings mound reached a critical height, slumping occurred, carrying tailings cross-inlet (ICM Annual Environmental Report 1972, isopach map of total mine-derived material), but tailings also started to creep down-inlet. At this time, the tailings pile was 6 m thick off the outfall, and another 6-m-deep pile had formed at the base of the south wall. Also at this time, one could see evidence of waste rock material being deposited on the inlet floor off the north side of Rupert Inlet.

By September 1973, the previously described depositional pattern continued (ICM Annual Environmental Report 1973 isopach map of total mine-derived material). The tailings pile off the outfall continued to expand, and the down-inlet transport of tailings accelerated. The underwater waste-rock pile expanded. In the near field, however, a channel developed on top of the tailings pile off the outfall, and a shallow seafloor valley started to form down-inlet. By November 1974, the Island Copper CSP survey revealed that a clear depositional pattern had emerged with a well-developed seafloor valley, carrying tailings down-inlet toward Hankin Point, and that the whole tailings pile and waste rock pile continued to grow. Johnson had studied this system before, and after mining operations began, he conducted further study for his Ph.D. thesis at UBC (Johnson 1974). He concluded that before mining began, during deep-water renewal, sediments were eroded from the deeper parts of the basin near Hankin Point and deposited on the northern slope of Rupert Inlet. He also made near-bottom current measurements after tailings discharge began and observed several short-lived events of high-amplitude (greater than 50 cm/sec) down-inlet currents. These, he suggested, were turbidity current surges.

Following Johnson (1974), Hay (1981, 1982, 1983) and Hay, Burling, and Murray (1982) published the results of their research on Rupert Inlet. Hay demonstrated conclusively from echo-sounding, high-resolution CSP and sediment cores that tailings were being deposited down-inlet. The source of sediment was quasi-steady as to the rate, type, and size of material discharged. Superimposed on this quasi-steady input, however, was the irregular occurrence of surge-type turbidity currents resulting from the slumping of previously deposited tailings. Hence continuous and surge-type turbidity currents are expected to be an important aspect of channel development in these systems.

The CSPs obtained over the years at Rupert Inlet show that from 1973 on, a leveed seafloor valley developed that clearly indicated the occurrence of flow channel overspill from both continuous and surge flows. Both phenomena were observed and measured by Hay (1982), and the channel system was reported by Hay, Murray, and Burling (1983).

The localized thickness of tailings downstream from the site of maximum deposition is the result of infilling of topographic depressions in the pre-mine bathymetry. Furthermore, Hay (1982) furnished additional data to the system from his detailed surveys, which better defined the position of the channel axis and the associated levees. He also showed a series of meanders in the channel immediately downstream from the bend opposite the outfall.

This leveed submarine channel system (Hay, Murray, and Burling 1983) continued to carry mine tailings down-inlet until mine closure in December 1995. This conclusion is substantiated further by the 1988 ICM CSP survey and by the yearly monitoring that included an echo-sounding survey used to track the movement and position of the submarine channel system and associated tailings.

Some of the finer fractions of the tailings solids were entrained into the ambient water and transported as suspended load (Hay, Murray, and Burling 1983). A small part of this suspended load was lost through Quatsino Narrows and off to Holberg Inlet. The finer fraction of the waste rock was subjected to similar episodes of dispersion and settlement. As discussed earlier, the coarser fraction of the waste rock was mostly confined within the area of dumping. The waste rock infilled the channel by particles sliding and rolling down slopes in contrast to turbidity current transport of the tailings. During this process some sorting of the waste rock must have taken place, with the coarser fraction reaching the bottom first and building the toe of the slope. In addition, periodic slope failures along the face of the growing marine waste rock pile carried surges of waste rock across Rupert Inlet and partway up the south slope. Some of these failures amounted to nearly 1 million tonnes of waste-rock movement.

Figure 4.10 presents a schematic depositional model of mine-related materials. The model explains the depositional processes discussed above. The support for the model can be found in Hay's delineation of channels (Hay 1982). He focused his discussion on the morphological description of one major channel, which has a meandering pattern in the inlet floor. Although the existence of other minor deltaic channels is obvious, he does not discuss them. The presence of a meandering pattern of the major channel on the inlet floor indicates the relative importance of friction over inertia.

Geomorphic and Bathymetric Changes in Rupert Inlet

A qualitative and quantitative assessment of the changes in the bathymetry of Rupert Inlet was conducted using the pre-mine and the January 2000 bathymetric surveys (Rescan 2000) as follows:

- Qualitatively, by superimposition of selected isobaths, by superimposition of some selected cross-sectional profiles, and through use of an isopach map prepared by obtaining the difference in vertical heights

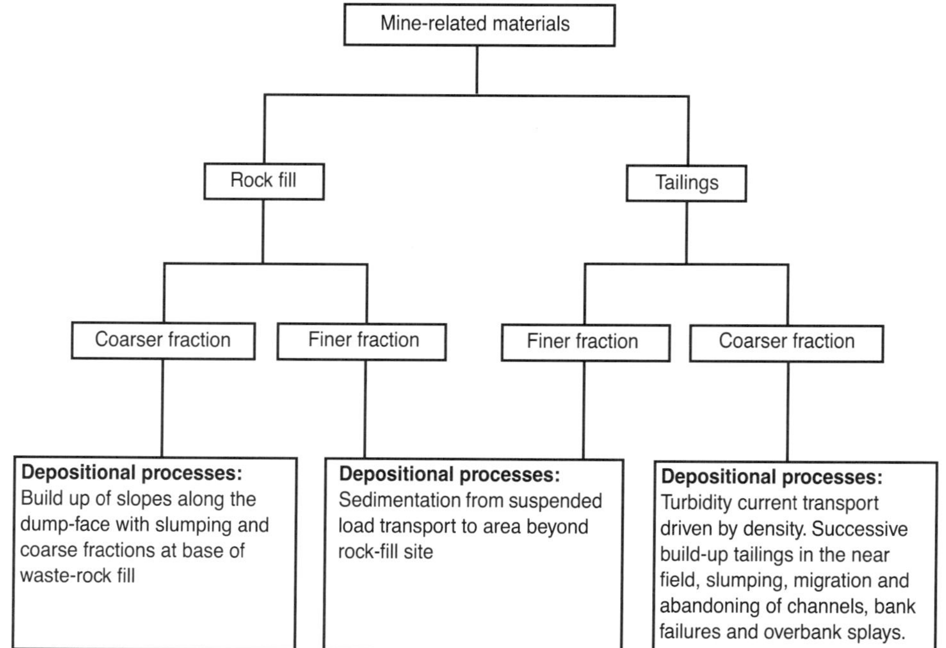

FIGURE 4.10   Schematic depositional model of the mine-related materials in Rupert/Holberg Inlet

- Quantitatively, by plotting of a hypsometric relation between volume or area as a function of depth for the whole inlet for different cross sections
- Quantitatively by estimating a mine-related material budget based on all available information

Visual Change

Figure 4.11 shows the superimposition of selected isobaths taken from the pre-mine and the January 2000 bathymetric surveys. There are two salient features to this superimposition. The first is an overall shift of the isobaths toward the west, indicating infilling of Rupert Inlet by mine-related sediments as previously described. The large, deep area covered by the 150-m isobath during the pre-mine period has virtually disappeared. The 90-m and the 120-m isobaths are shifted westward by more than 6 km toward Holberg Inlet. The second is the reorientation of the 30- and 60-m isobaths in the waste-rock area.

Figures 4.12 through 4.16 show the superimposed cross-sectional profiles of the pre-mine and January 2000 bathymetric surveys. The effects of the waste rock can be seen in the superimposed profiles of cross sections 1 through 3. At cross section 2, the shoreline advanced southward by more than 1 km in response to waste rock.

Figure 4.17 shows the isopach map that is obtained by estimating the difference between the pre-mine and the January 2000 bathymetric charts. The contours represent the deposition thickness isolines for the entire area of Rupert/Holberg Inlet. A pattern of deposition is dominated by two mechanisms of mine-related material disposal. These are the shoreline waste-rock area and the tailings placement in Rupert Inlet. This isopach map is similar to the isopach map prepared from the January 2000 seismic survey (Figure 4.18).

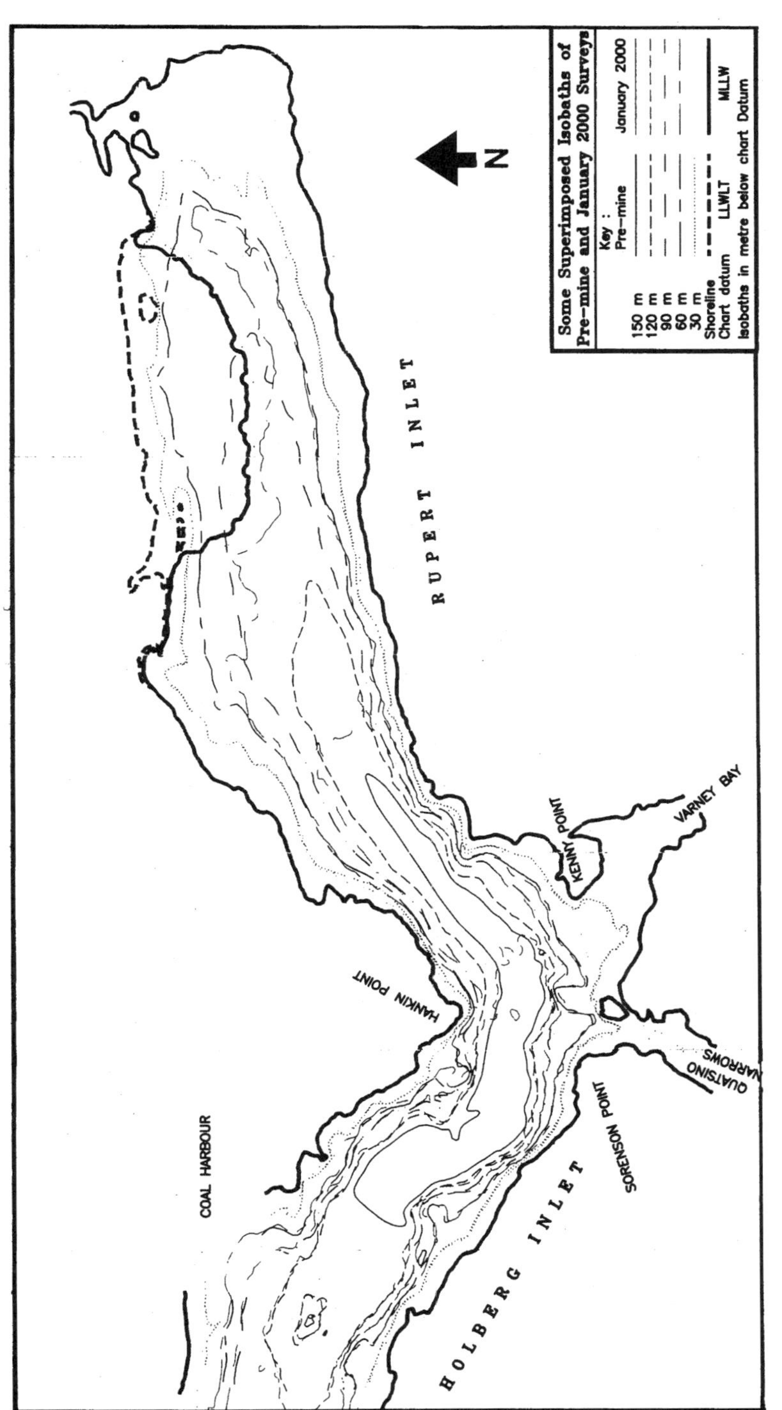

FIGURE 4.11   Superimposed isobaths of pre-mine and January 2000 surveys

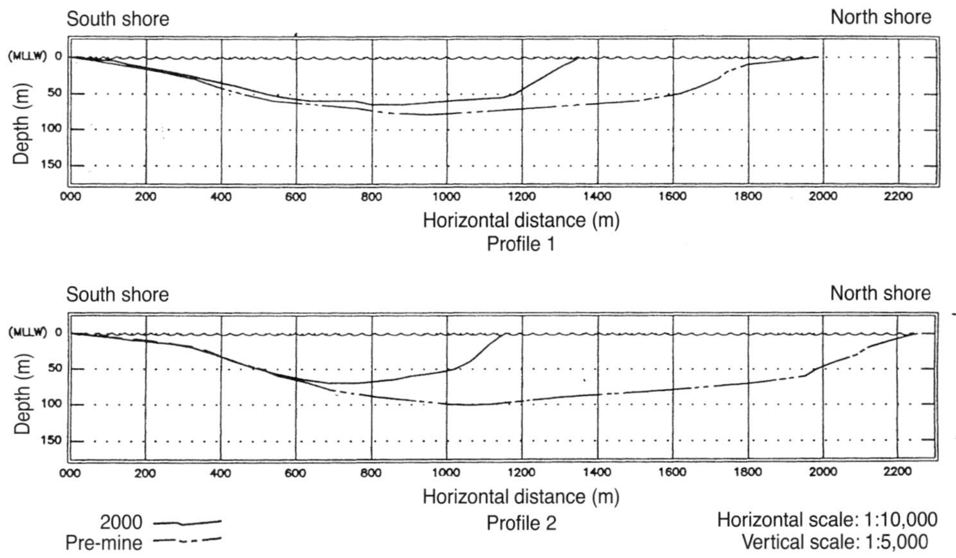

FIGURE 4.12   Superimposed profiles of cross sections 1 and 2

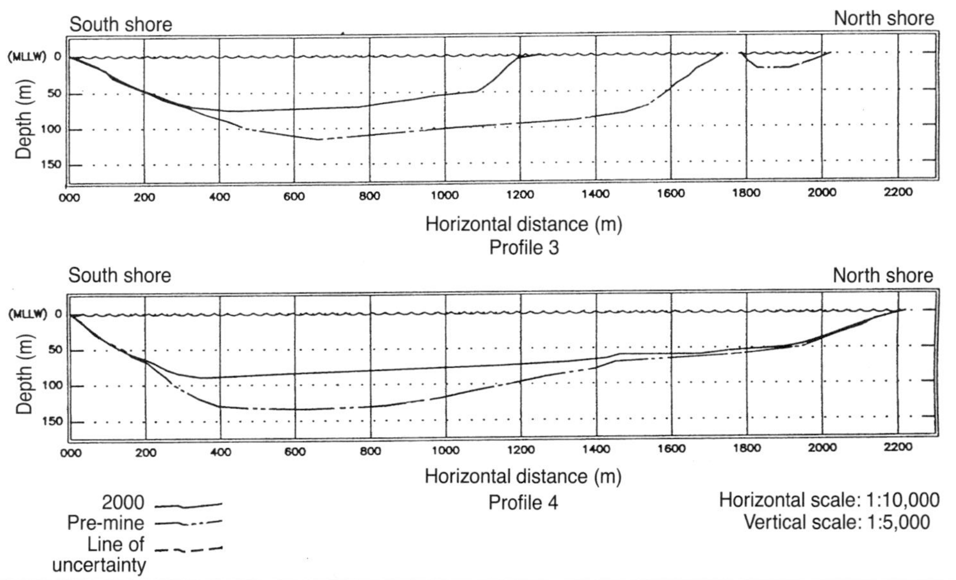

FIGURE 4.13   Superimposed profiles of cross sections 3 and 4

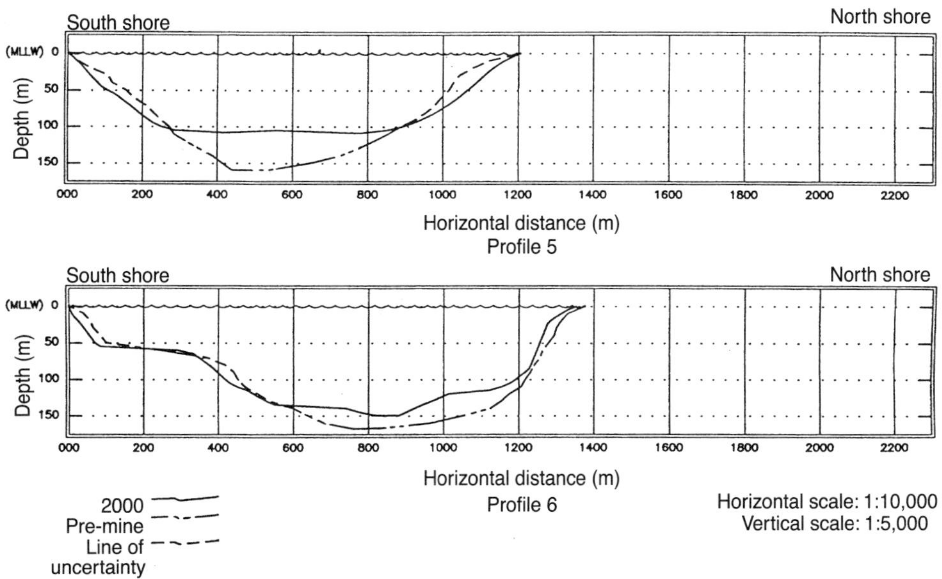

FIGURE 4.14 Superimposed profiles of cross sections 5 and 6

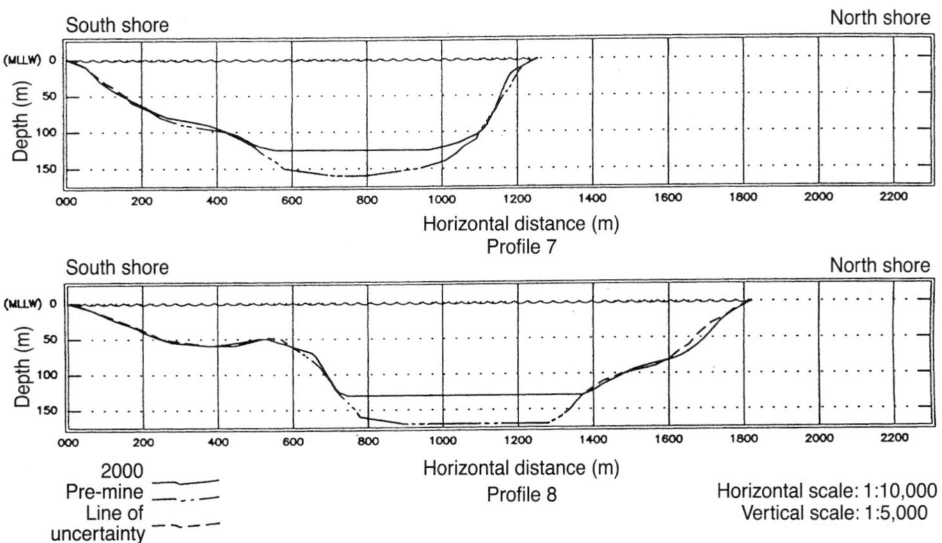

FIGURE 4.15 Superimposed profiles of cross sections 7 and 8

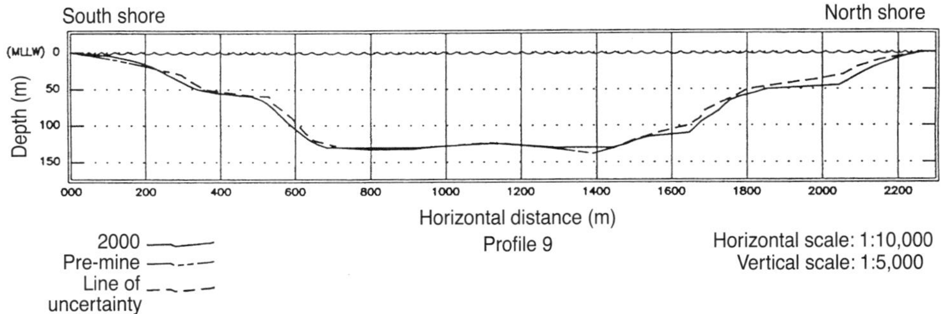

FIGURE 4.16 Superimposed profiles of cross section 9

Turbidity Problems at the Confluence Scour Hole at Hankin Point

A scour hole exists off Hankin Point at the confluence of Quatsino Narrows, Rupert Inlet, and Holberg Inlet (Figures 4.2 and 4.9). Existence of a confluence scour hole is common at the point of convergence (Bridge 1993). Such scour holes are caused by and sustained on a mobile bed by erosive powers of turbulence and eddies generated at the confluence. In this situation, the scouring is dominated by strong, density-driven currents from Quatsino Narrows during incoming tides. Because of west coast upwelling in spring and summer, relatively cold and dense waters enter through Quatsino Narrows during incoming tide (Pederson et al. 1993). Additionally, the freshwater discharge from the Marble River causes the inlet water to be of lower density than the incoming water through Quatsino Narrows. Assuming that the density of incoming water is sufficiently higher than the resident water, the incoming flow plunges downward with the formation of a counterclockwise gyre. This gyre can be visualized as a counterclockwise motion on a vertical plane across the channel, causing erosion of the hole.

The nature and strength of the circulation and scouring are not constant in time and depend on:

- Density contrast between the incoming and resident water
- Phase of the tidal cycle—small, large, neap, and spring tides—as well as incoming and outgoing tides
- Availability of erodible material

Figure 4.19 shows a seismic section through the scour hole along line X7 (see also Figure 4.6). Different failure structures are noticeable on the slopes of the hole with an eroded escarpment on the east slope (Rupert Inlet). The western slope (Holberg Inlet) of the hole is about 8°. The deepest point of the hole has a depth of about 160 m. At this location, the pre-mine bathymetry (Figure 4.2) shows a scour hole surrounded by the 170-m isobath, with the deepest point at 180 m. The difference between the two gives the year 2000 thickness of deposits at the deepest point (20 m). Figure 4.20 shows the year 2000 thickness and 20 years (1976–1996) of quarterly time-series measurements of deposit thickness reported by ICM (1996). The deposits show the gradual infilling of the hole as tailings continued to flow from the mine until December 1995. The figure shows that in the 4 years from the date of mine closure in December 1995 to the January 2000 measurement date, 33 m of deposits were scoured out. If scouring continued at this rate, it would take about 3 additional years to scour the deposits to the original bed.

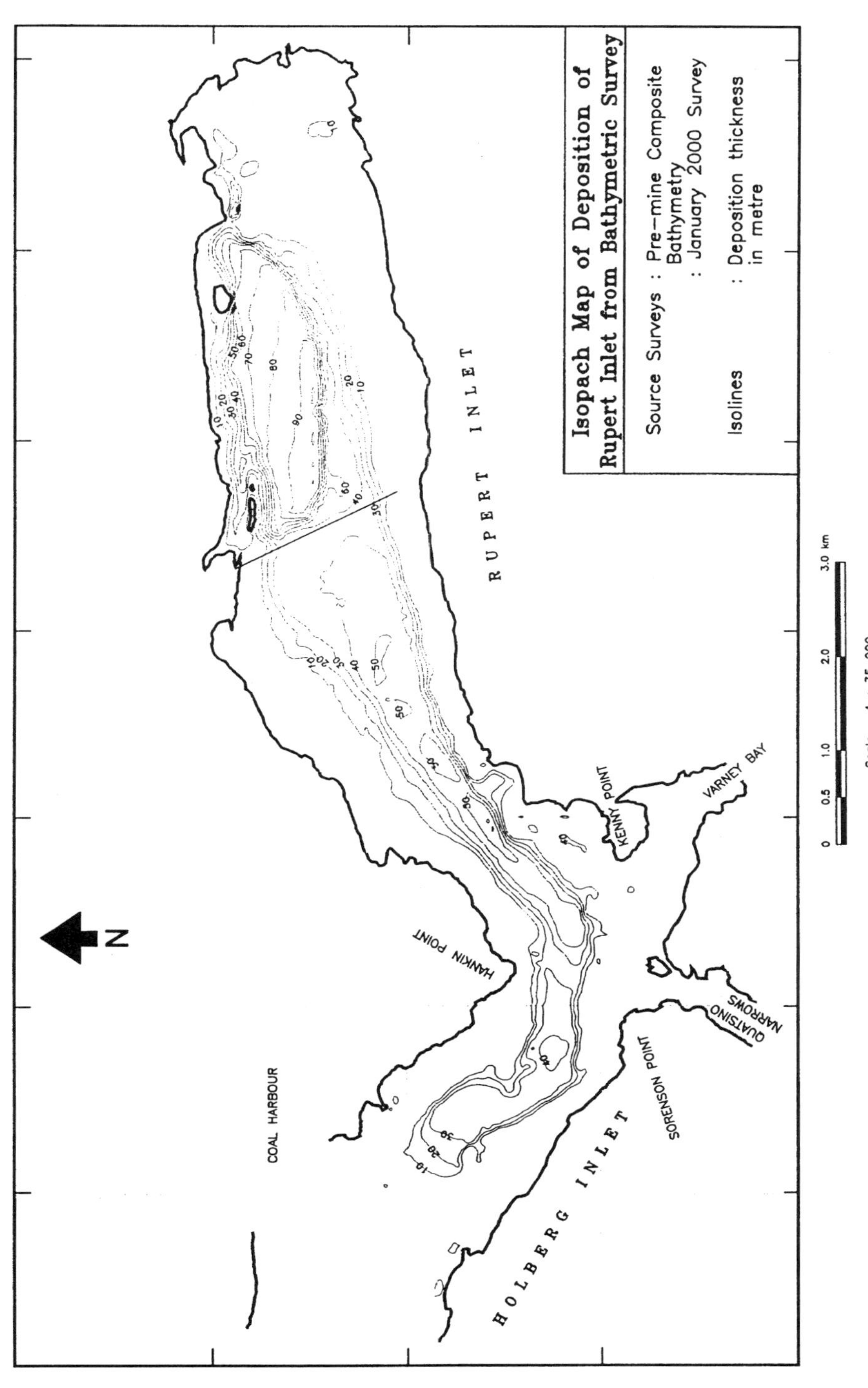

FIGURE 4.17 Isopach map of deposition of Rupert Inlet from bathymetric survey

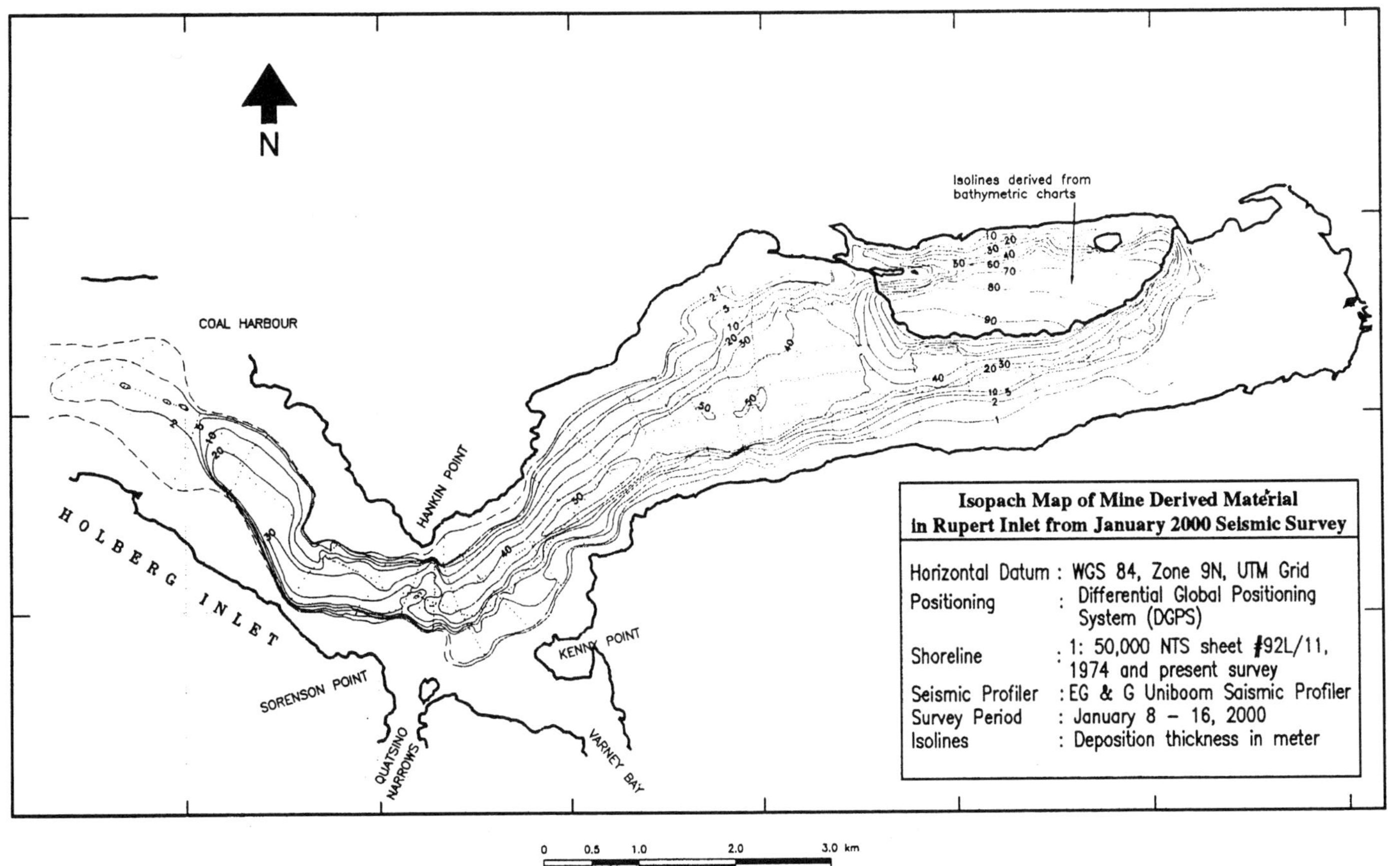

FIGURE 4.18   Isopach map of mine-derived material in Rupert Inlet from January 2000 seismic survey

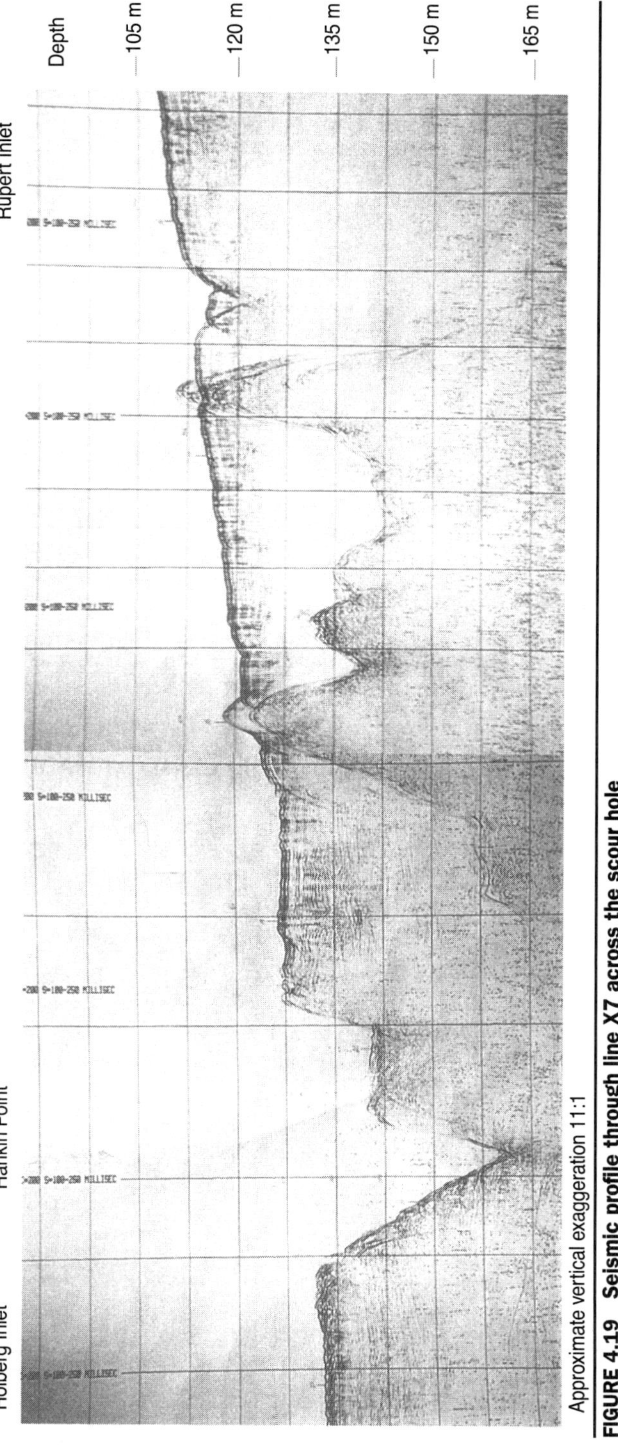

FIGURE 4.19 Seismic profile through line X7 across the scour hole

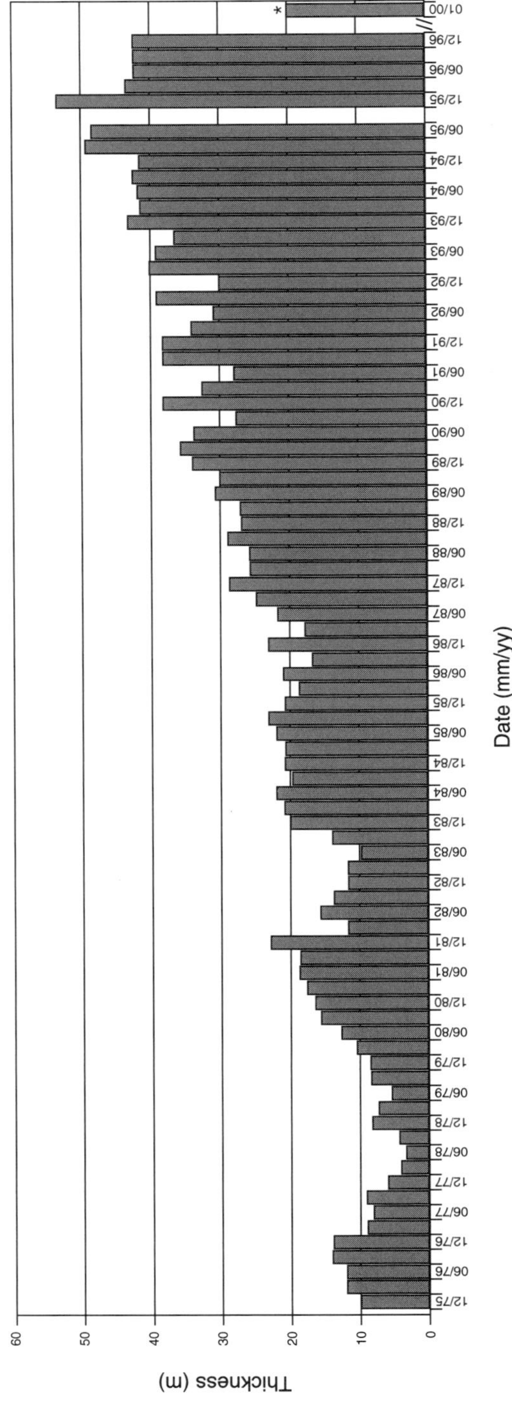

*Thickness at the deepest point of the scour pit based on January 2000 seismic survey

FIGURE 4.20   Thickness of deposits in the confluence scour hole at Hankin Point (1976–2000)

Hypsometry

Figures 4.21 through 4.26 present the channel hypsometry for the whole inlet and for a number of cross sections. The pre-mine and the January 2000 hypsometric curves are superimposed to show the channel volume changes brought about by deposition. Figure 4.21 represents the entire Rupert Inlet with a portion of Holberg Inlet, whereas Figure 4.4 showed the boundaries of the inlet used for estimating the volume. The western boundary is formed by cross section 9. The east–west line a–b forms the southern boundary through the Quatsino Narrows. These boundary lines represent the approximate limits of the bathymetric survey. Beyond these limits, the sounding tracks are not dense enough for accurate generation of isobaths. Comparison of the pre-mine and the January 2000 data indicates that deposition occurred at all depths, although most of the tailings are now in areas deeper than 50 m.

The shoreline rock dump causes the deposition in depths shallower than 40 m. The pre-mine inlet volume (for the area shown in Figure 4.4) was 1.81×10^9 m^3 compared with the January 2000 volume of 1.36×10^9 m^3. This gives a net deposition of about 0.45×10^9 m^3. The effects of shoreline waste rock are most noticeable in cross sections 1, 2, and 3 (Figures 4.22 [a,b] and 4.23a). They show deposition at all depths. Westward from cross section 4 (Figure 4.23b), the effects of the shoreline waste rock cannot be seen in shallower water. At cross section 4, the tailings infilled only areas deeper than 60 m. Cross section 5 (Figure 4.24a) indicates some erratic changes, which are also evident from the superimposed profiles of the cross sections (Figure 4.15). At cross section 6 (Figure 4.24b), which represents the area at the confluence of Rupert and Holberg Inlets adjacent to Quatsino Narrows, the infilling occurs in areas deeper than about 120 m. Cross section 7 (Figure 4.25a) shows a similar picture with deposition occurring in areas

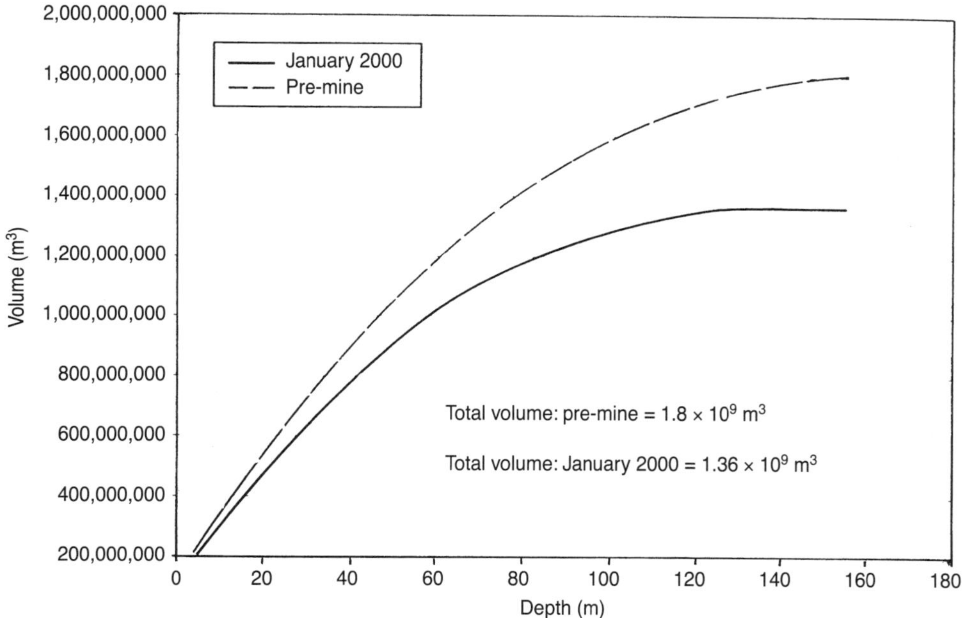

FIGURE 4.21 Pre-mine and present hypsometry of Rupert Inlet

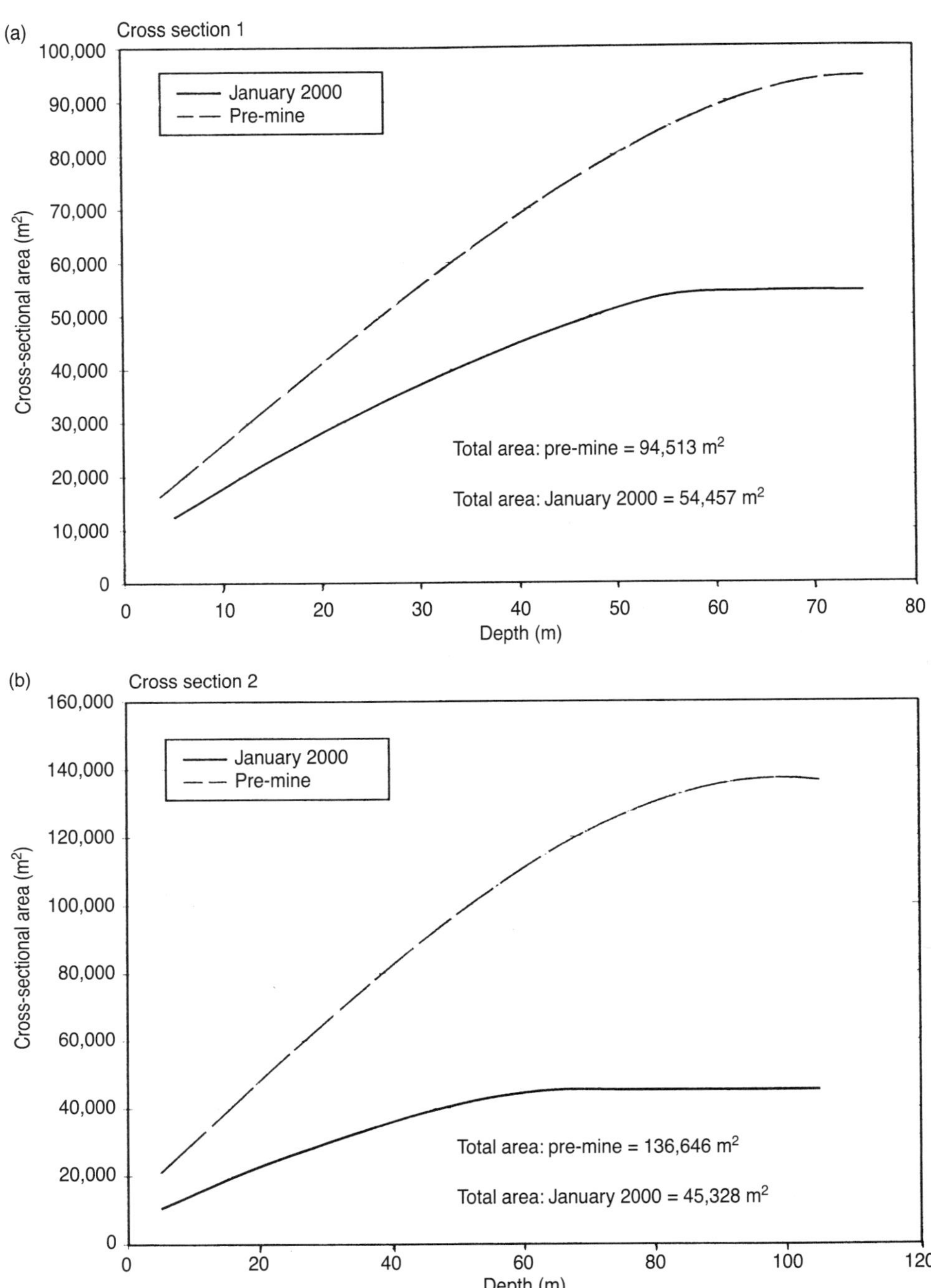

FIGURE 4.22 **Pre-mine and present hypsometry of Rupert Inlet at cross sections 1 and 2**

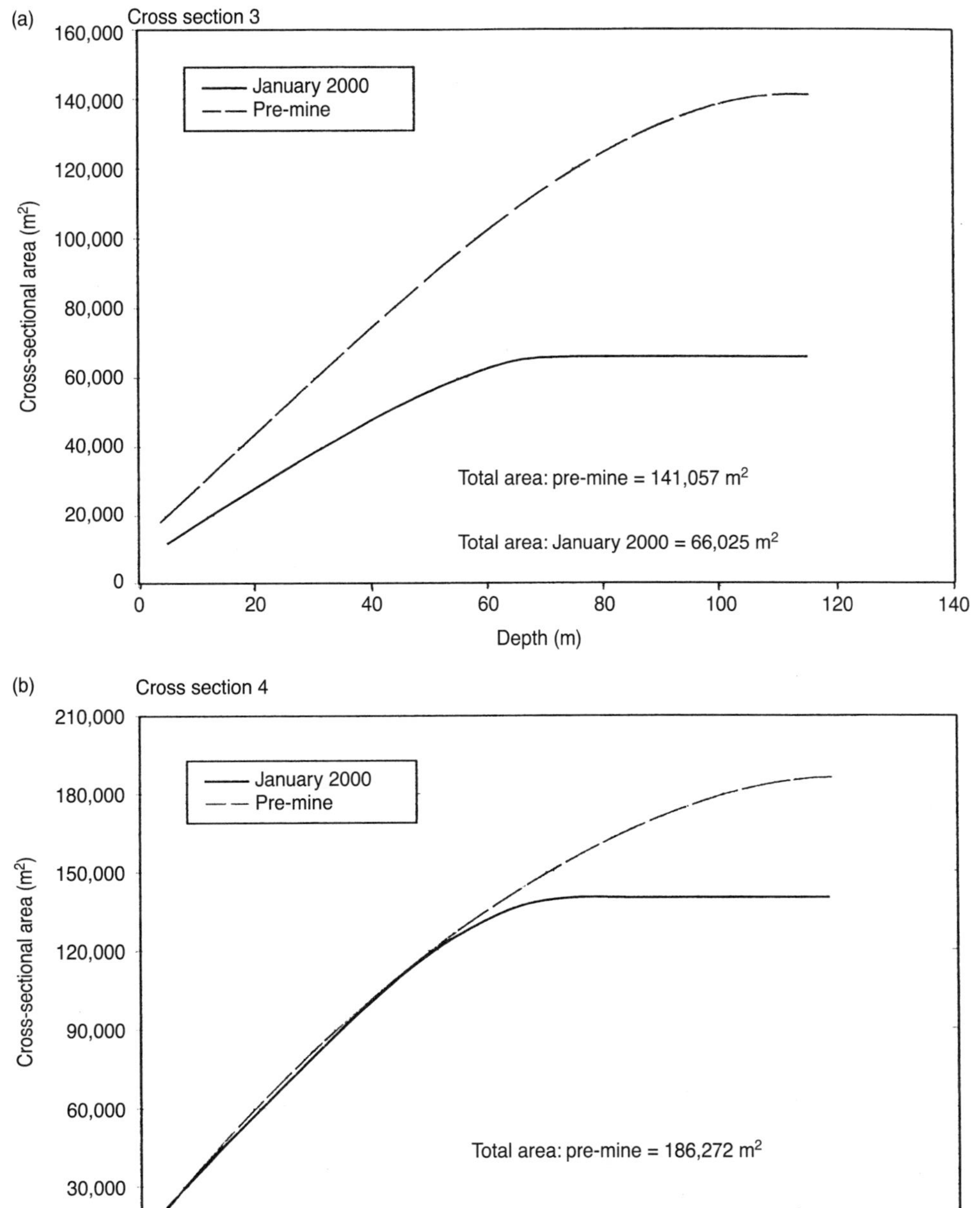

FIGURE 4.23 **Pre-mine and present hypsometry of Rupert Inlet at cross sections 3 and 4**

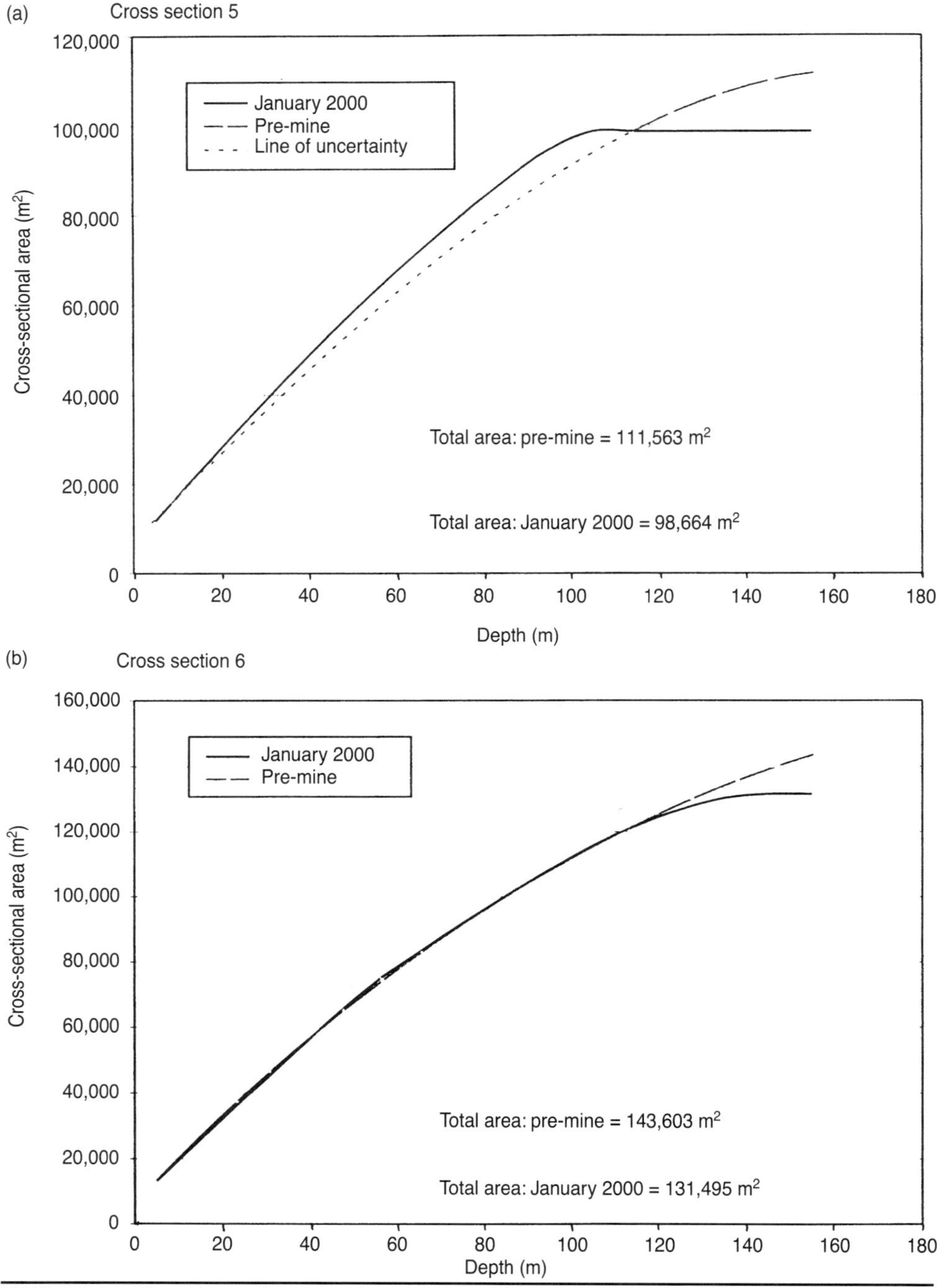

FIGURE 4.24 **Pre-mine and present hypsometry of Rupert Inlet at cross sections 5 and 6**

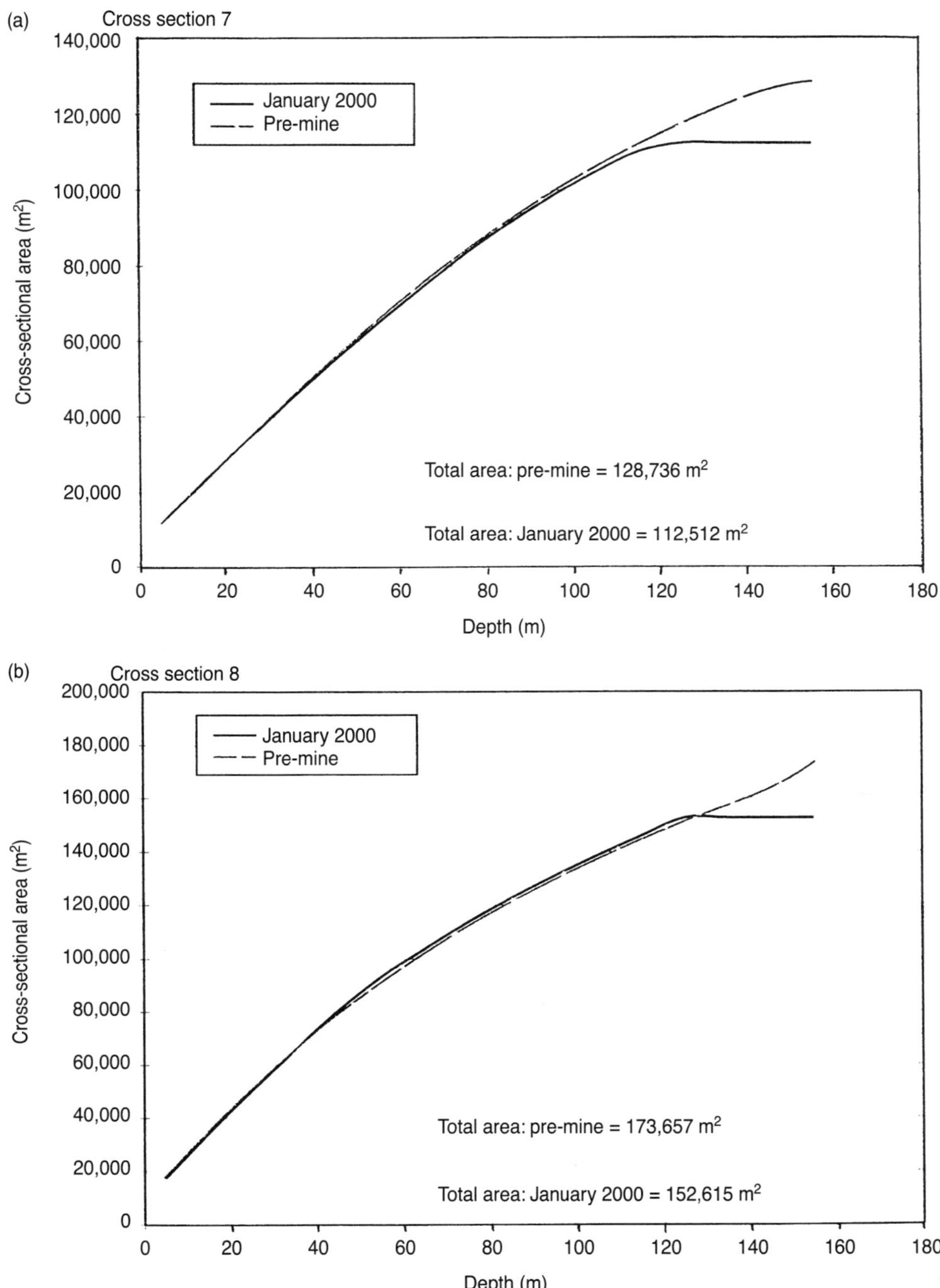

FIGURE 4.25   Pre-mine and present hypsometry of Rupert Inlet at cross sections 7 and 8

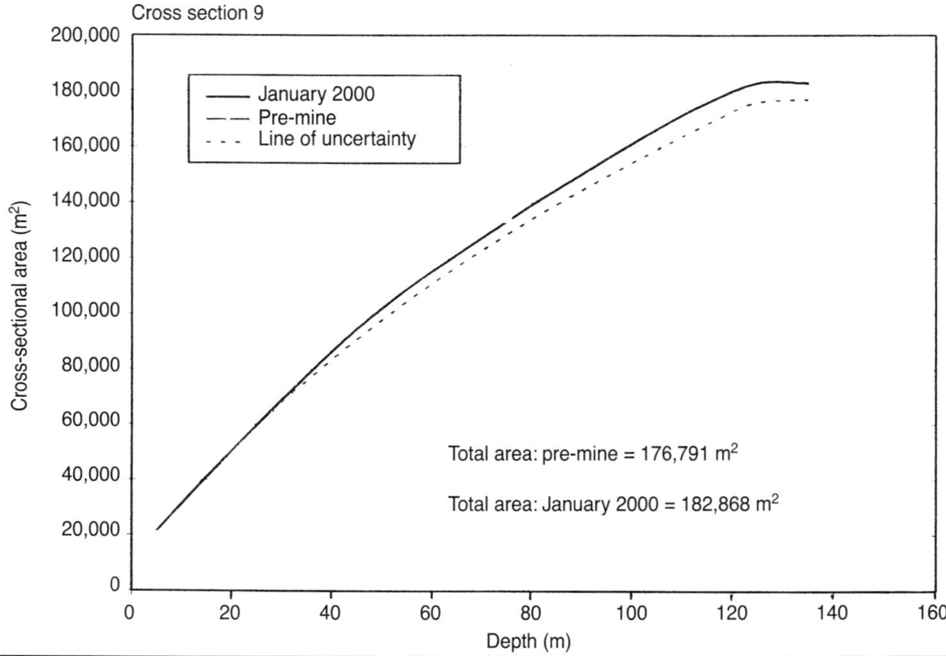

FIGURE 4.26 Pre-mine and present hypsometry of Rupert Inlet at cross section 9

deeper than 120 m. At cross section 8 (Figure 4.25b), the pre-mine hypsometry shows an erratic trend with the slope bending upward. Cross section 9 (Figure 4.26) shows an unrealistic picture of channel erosion, which is indicative of a discrepancy between the original survey and the more accurate January 2000 survey.

Mine-Related Material Deposition Budget

A quantitative analysis of changes in Rupert/Holberg Inlet morphology and a mine-material budget can be made based on comparison of the pre-mine and the January 2000 bathymetric surveys. Table 4.3 summarizes changes in channel volume and cross-sectional areas. Overall, 24.9% of Rupert Inlet volume has been filled in by shoreline rock placement and tailings placement. The cross-sectional area comparison shows that most of the deposition has occurred in Rupert Inlet. At cross sections 1 to 3, the inlet is filled in by more than 50% because of the shoreline rock placement. Cross section 4 (Figure 4.23b), located immediately west of the marine outfall terminus, shows the largest effect of tailings deposition. At cross section 6 (Figure 4.24b), which passes through the confluence scour hole off the Quatsino Narrows, the effects of tailings deposition are much less as one proceeds into Holberg Inlet. Off the town of Coal Harbour, any small effect of deposition is below the resolution of the bathymetric survey.

When preparing the solids budget, various sources and receptors (sinks) of the mine-related material were taken into account.

TABLE 4.3 Changes in volume/cross-sectional area of Rupert/Holberg Inlet as a result of mine-related material disposal

Description	Pre-mine	January 2000	Change	Infill by Deposition (%)
Total volume (m^3)	1.81×10^9	1.36×10^9	0.45×10^9	24.9
Area for cross section 1 (m^2)	94,513	54,457	40,056	42.4
Area for cross section 2 (m^2)	136,646	45,328	91,318	66.8
Area for cross section 3 (m^2)	141,057	66,025	75,032	53.2
Area for cross section 4 (m^2)	186,272	140,125	46,147	24.8
Area for cross section 5 (m^2)	111,563	98,664	12,899	11.6
Area for cross section 6 (m^2)	143,603	131,495	12,108	8.4
Area for cross section 7 (m^2)	128,736	112,512	16,224	12.6
Area for cross section 8 (m^2)	173,657	152,615	21,042	12.1
Area for cross section 9 (m^2)	176,791	182,868	–6,007	–3.4* (~0)

*Differences among bathymetric surveys along the inlet rocky walls are unrealistic and result from inaccuracies of soundings on steep slopes.

Sources

- Shoreline waste rock at depths below datum
- Tailings

Receptors

- Deposition in Rupert/Holberg Inlet
- Loss through Quatsino Narrows (beyond section a–b, Figure 4.4)
- Loss to Holberg Inlet (west of Coal Harbour—beyond cross section 9, Figure 4.4)

The estimation of these sources, depositions, and losses in tonnage were made below a datum, which was taken as the shoreline (or 0-line), derived from a 92L/11 NTS sheet (see, for example, Figure 4.2). All these sources, depositions, and losses are evaluated separately. Before going further, a realistic estimate of the dry density of the waste materials was necessary. Among variables, dry density depends on the grain-size characteristics and porosity of the sediments. The two basic sources of materials—the shoreline waste rock and the tailings—have widely differing particle size distributions. The waste rock consisted of drilled and blasted rock with some sizes as large as 1 m. It was subjected to mechanical compaction after being dumped along the shore face. The tailings, on the other hand, were fine-textured with lower porosity. The tailings had a volumetric concentration of 10% and a median particle diameter of 30 µm (Evans and Poling 1975).

The densities of these materials have been estimated based on a number of earlier observations. Assumed to be dominated by quartz minerals, the rocks derived from the Island Copper orebody have a granular density of 2.65 g/cm^3. With an unsorted sandy material porosity of 33%, the dry density of materials dominated by quartz grain would be 1,776 kg/m^3. The *Acid Mine Drainage Study of the North Dump, Island Copper Mine* (UBC 1991) showed that the average dry density of the dump material was 1,847 kg/m^3. This value is accepted for the estimation of the waste-rock tonnage.

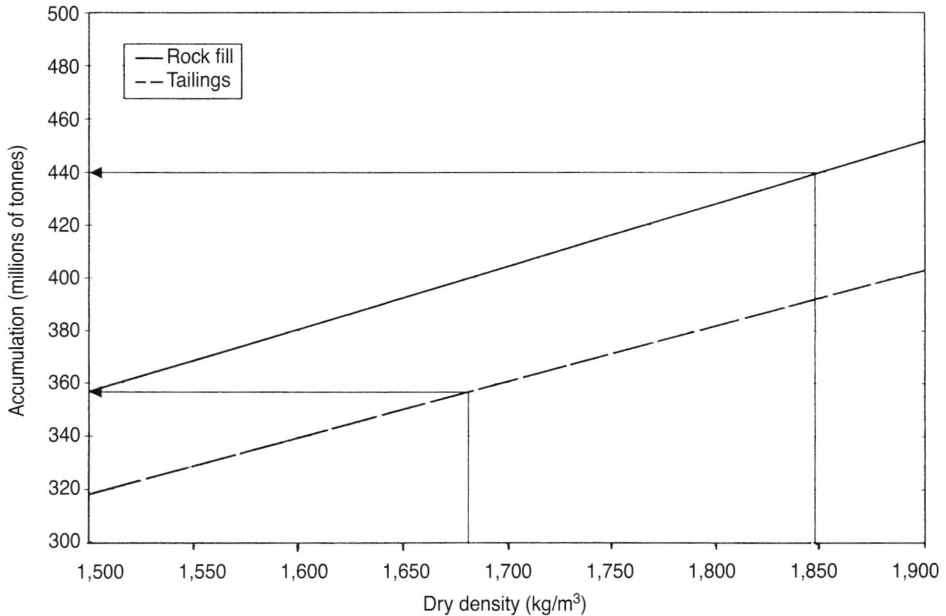

FIGURE 4.27   Sensitivity of rock fill and tailings tonnage to change in dry density

The tailings, which consisted of finer particles, consolidated slowly and had a lower dry density than the waste-rock material. However, this is not true for the entire deposition. Horizontal sorting during transport and differential consolidation over the vertical area may result in a widely varying dry density. The seismic profiling of some areas indicated that mill tailings still have a high water content. A vertical stratification occurs with an increase in consolidation with depth of tailings. Failure in the past to get a sample in some places by an underwater sampler suggests that horizontal sorting has caused a segregated accumulation of fine and coarse materials in the inlet. An average representative dry density is required to estimate the tonnage. The British Columbia Department of Energy, Mines and Petroleum has determined that tailings have a dry density varying from 1,603 to 1,763 kg/m³ (100 to 110 lb/ft³). An average dry density of 1,683 kg/m³ (105 lb/ft³), which is equivalent to 1,682 kg/m³, was used.

For this analysis, 1,847 and 1,682 kg/m³ of dry densities are assumed for the waste rock and tailings, respectively. Figure 4.27 presents the sensitivity of the estimated tonnage to changes in dry density of the waste rock and tailings.

Waste rock in depths below datum. According to the ICM *Reclamation Report* (ICM 1999), an area of 262 ha of new land has been reclaimed as a flat surface along the north shore of the Rupert Inlet. It is estimated that some 560 million tonnes of rock were placed along the north shore of Rupert Inlet. Of this amount, some waste rock occupies areas above the datum from the total. According to a map prepared by ICM, the waste rock extends some 8 m above datum. For a reclaimed area of 262 ha, this would mean a volume of $20.96 \times 10^6 \text{ m}^3$ above datum. With an assumed dry density of 1,847.7 kg/m³, this translates to 39 million tonnes. Subtracting this number from the total gives the waste rock below datum at 521 million tonnes.

Estimated tailings discharge. The ICM *Reclamation Report* (ICM 1999) estimates that mill feed was 363 million tonnes. It is estimated that some 5 million tonnes of metal sulfide concentrates were produced from that mill feed. The difference between the two gives the total mill tailings discharged at 358 million tonnes.

Estimated deposition in the Rupert Inlet. As discussed earlier, total volume deposited in Rupert Inlet was 0.45×10^9 m^3. This deposition must be separated according to its sources. Some assumptions are necessary to make a reasonable separation of volume. The finer fraction of waste rock was transported to outside the disposal area, but the tidal current redistributed the fine tailings to different areas, including the waste-rock zones. The finer fraction of the waste rock, which could escape the dump zone, is assumed to be a very negligible portion of the total dump. Therefore, it can be assumed that waste rock is mostly concentrated within the zone of deposition. The isopach map drawn in Figure 4.10 is used to estimate the waste rock below datum. Areas covered by different isobaths were estimated and then multiplied by the average thickness of the deposition. A total beach deposition of 237.85×10^6 m^3 has been calculated. For a dry density of 1,847.4 kg/m^3, this represents 440 million tonnes of waste-rock dump below datum.

The remaining deposition volume accounting for the tailings is 212.15×10^6 m^3, which represents 357 million tonnes for a dry density of 1,682 kg/m^3. A comparison of the estimated tailings volume with the January 2000 inlet volume (Table 4.3) shows that tailings have occupied 15.6% of the two inlets (up to section 9 of Holberg Inlet as shown in Figure 4.4). Utah Mining (1969) predicted a 10% infilling of the two inlets in total by tailings. This, as it turns out, was quite accurate.

The total of these two estimated mine rock sources in Rupert/Holberg Inlet is 797 million tonnes.

Estimated loss through Quatsino Narrows. In July 1980, ICM's Environmental Department conducted a 24-hour, suspended-solid survey in Quatsino Narrows. Based on those measurements, one can estimate the total sediment loss for the 25 years of operations (1971–1995) at 1.4 million tonnes. This represents 0.4% of the tailings discharged to the inlets.

Loss to Holberg Inlet (west of Coal Harbour). There are no measurements to estimate the tonnage of mill tailings that deposited beyond the entrance to Coal Harbour. As discussed earlier, the hypsometry shows no significant deposition beyond Coal Harbour. Some loss of finer tailings is expected, and following the measurements at Quatsino Narrows, the loss is probably in the order of 0.4% or 1.4 million tonnes.

Uncertainty of estimates. It is difficult to balance several quantities, especially when using measurements from different sources. In this case, uncertainties result from measurements, map preparation, and estimation of volumes and densities. Usually the measurements of lengths (but in this case, length, width, and depth) have an uncertainty of ±5%, and measurements of density have an uncertainty of ±2% (Soulby 1977). This would give an uncertainty, arising only from measurements, of about ±10%. The map preparation and estimation volume may contribute another ±5%, giving a total uncertainty of ±15%.

For all estimates based on bathymetry, an uncertainty of ±15% applies. Table 4.4 summarizes all the estimates. The overall distribution of the waste rock and budget gives

TABLE 4.4 Summary of mine-related material budget

Item	Estimated Volume (million m³)	Estimated Tonnage (million tonnes)
Estimated volume of tonnage of rock fill		
ICM *Reclamation Report* (ICM 1999)—total placed	—	560.0
Estimated fraction above datum	31.0	–39.0 ± 15%
Rock fill below datum	—	521.0
Estimated input through tailings		
ICM *Reclamation Report* (ICM 1999)—mill feed	—	363.0
Approximate abstraction of metal concentrates	—	–5.0
Tailings through submarine outfall	—	358.0
Estimated deposition from comparisons of bathymetry	450.0	
Waste rock (estimated from isopach map)	237.8	440.0 ± 15%
Tailings	212.2	357.0 ± 15%
Estimated loss through Quatsino Narrows	—	1.4 ± 15%
Estimated loss to Holberg Inlet	—	~1.4 ± 15%
Total input through rock fill and tailings	—	879.0
Total deposition and losses	—	800.0 ± 15%
Discrepancy in total estimate (percentage of input)	—	9.0%
Discrepancy in tailings (percentage of input)	—	0.3%

Notes:

    Assumed dry density, waste-rock materials: 1,874.4 kg/m³.

    Assumed dry density, tailings: 1,682.0 kg/m³.

a reasonable picture of morphological changes in Rupert Inlet. All the inputs of waste rock and tailings could be accounted for by estimating the deposition in Rupert Inlet.

Total balance. The total deposition, including losses, is estimated at 800 million tonnes against the best-estimated input of 879 million tonnes. The discrepancy between the two is about 79 million tonnes or 9% of the input. The discrepancy is better for the budget of the tailings alone. For the 358 million tonnes of tailings actually discharged, 357 million tonnes of deposition could be accounted for. The discrepancy is only 1 million tonnes or 0.3% of the input. This close agreement may be more fortuitous than accurate.

Sensitivity of estimated tonnage to dry density. The estimation of tonnage from volumes depends largely on the selected dry density. Because there are no direct dry density measurements of either the waste rock or the tailings, some reasonable values derived from other areas were used. Figure 4.27 illustrates the sensitivity of tonnage to changes in dry density. The selected density and the estimates are shown for the waste rock and the tailings. It is obvious that different estimates of tonnage can be predicted depending on the density selected.

Rescan (1999) has completed numerous consolidation tests of tailing samples for another project. It was found that the dry density of tailings, characterized by a median diameter of 25 μm, varied from 1,250 kg/m³ under its own pressure to 1,580 kg/m³ under a 4-m cover of tailings. For a sample with a median diameter of 120 μm, the dry density varied from 1,560 kg/m^3 under its own pressure to 1,810 kg/m^3 under a 4-m cover of tailings.

SUMMARY AND CONCLUSIONS

This chapter integrates the results of numerous bathymetric and CSP surveys and many other scientific studies conducted in Rupert Inlet over a 29-year period. It also makes an assessment of morphological changes caused by mine-related infilling.

The findings are reported qualitatively and quantitatively through comparisons of isobaths and isopach maps, superimposition of some selected profiles, hypsometric relations, and estimated volumes. A mine-related materials budget is presented by taking into account all the sources, depositions, and losses in receptors. The findings are:

- The bathymetry of Rupert Inlet changed considerably following the disposal of mine-related infilling. Isobaths from the year 2000 show the existence of channels on the inlet seafloor resulting from bed-load transport and redistribution of mine-derived material. These channels represent remnants of earlier turbidity current transport. With no tailings deposition since the end of December 1995, the recent seafloor reworking has been a result of the shearing effects of tidal currents, in contrast to turbidity current processes during the active disposal period. The depositional processes occurred as bed-load transport, driven by density currents, with successive buildup of tailings, sorting of materials, shifting and overtopping of channels, bank failure, and bank splays.

- The comparison of isobaths shows that the 90-m and deeper isobaths shifted west into Holberg Inlet by more than 6 km.

- The isopach map shows higher deposition of tailings in deeper areas. Waste rock deposition can be seen clearly at the head of the inlet. In general, tailings deposition stayed below 50–60 m depth. The depositional processes of waste rock occurred as shoreward progression along the dump-face with the coarser fraction reaching the bottom and building up the base.

- Superimposition of nine selected cross-sectional profiles shows that the effects of the waste rock on the north shore are primarily concentrated in cross sections 1 to 3. This has also been shown by hypsometric curves at these three cross sections. In other places, the effects of mine-related material deposition are primarily seen in deeper areas. At the westerly cross section 9, near Coal Harbour, minimal deposition was seen.

- Estimates of volume for the area shown in Figure 4.4 and the cross sections show different rates of deposition on the inlet. To highlight this depositional pattern, some hypsometric curves are presented. Overall, the mine-related deposition has reduced the inlet volume by 25%. Near the waste rock areas, the cross sections are reduced by about 54%. The anthropogenic deposition occupies about 11% of the Rupert/Holberg system, which is close to the original estimate of 10% predicted in the brief presented by Utah Mining at the Public Inquiry in January 1971 to the British Columbia Pollution Control Board. Considering that, to date, the seafloor sediments have not reached static consolidation status, the figure is close.

- By taking into account all the possible inputs, depositions, and losses, a budget of mine-related material discharged into the inlet was made. By assuming some reasonable values of dry densities, the tonnage of waste rock material and the tailings were estimated from the volumes derived by comparing the bathymetric maps. Best-input estimates of mine-related material show that some 521 and 358 million tonnes were deposited as waste rock and tailings, respectively. The respective estimates from bathymetry are 440 and 357 million tonnes. An estimate shows that 0.4% or some 1.4 million tonnes were lost through Quatsino Narrows with a similar amount estimated to be deposited in Holberg Inlet. In summary, the total input into Rupert Inlet was 879 million tonnes, of which 800 million tonnes could be accounted for by deposition and losses. The discrepancy is 9% of the input. The estimated discrepancy for tailings is only 0.3%.

- It is noted that estimates of this nature suffer from large uncertainties. For these estimates, an uncertainty of ±15% applies.

ACKNOWLEDGMENTS

This chapter draws heavily upon the fieldwork and research conducted at the mine site by the authors, UBC graduate students, and other colleagues, including ICM environmental staff, during the past 30 years. We are particularly indebted to Robert D. MacDonald of the Geological Survey of Canada, whose fundamental fieldwork and data interpretation have made much of this summary possible. Similar contributions were made by former graduate students Alex Hay and Ron Johnson as well as professional colleagues Ron Burling, George Poling, Tim Parsons, and Derek Ellis. Finally, much of this summary chapter was made possible by the outstanding research and dedication of Rescan Environmental Services Ltd.

REFERENCES

BHP Billiton–Utah Mines Ltd. 1990. *Closure Plan.* Port Hardy, BC: BHP Billiton–Utah Mines Ltd.

BHP Billiton Minerals Canada Ltd. 1995. *Annual Environmental Assessment Report, Volume II.* Port Hardy, BC: ICM, BHP Billiton Minerals Ltd.

Bridge, J.S. 1993. The interaction between channel geometry, water flow, sediment transport and deposition in braided rivers. In *Braided Rivers,* J.L Best and C.S. Bristow, eds. Geological Society Special Publication 75:13–71.

Carstens, T., and E. Tesaker. 1972. Erosion by artificial suspension currents. Paper presented at the 13th Coastal Engineering Conference. Vancouver, BC. 4 pp.

CHS. 1999. *Tidal Manual.* Ottawa, Ontario: CHS. 25 pp.

Department of Energy, Mines and Resources. 1973. *Tentative Design Guide for Mine Waste Embankments in Canada.* Technical Bulletin TB 145. Ottawa, Ontario: Department of Energy, Mines, and Resources.

Drinkwater, K.F., and T.R. Osborn. 1975. The role of tidal mixing in Rupert and Holberg Inlets, Vancouver Island. *Limnol. Oceanogr.* 20:518–528.

Evans, J.B., and G.W. Poling. 1975. Discharging flotation mill tailings into a coastal inlet. *Proceedings of the 77th Annual Meeting of the Canadian Institute of Mining and Metallurgy.* Vancouver, BC.

Hay, A.E. 1981. Submarine channel formation and acoustic remote sensing of suspended sediments and turbidity current in Rupert Inlet, BC. Unpublished Ph.D. diss. UBC, Vancouver. 326 pp.

———. 1982. The effects of submarine channels on mine tailing disposal in Rupert Inlet, BC. In *Mine Tailings Disposal,* D.V. Ellis, ed. Ann Arbor, MI: Ann Arbor Science. 368 pp.

———. 1983. On the remote acoustic detection of suspended sediment at long wavelengths. *J. Geophys. Res.* 88:7525–7542.

Hay, A.E., R.W. Burling, and J.W. Murray. 1982. Remote acoustic detection of a turbidity current surge. *Science* 217:833–835.

Hay, A.E., J.W. Murray, and R.W. Burling. 1983. Submarine channels in Rupert Inlet, British Columbia: I. Morphology. *Sedimentary Geology* 36:269–315.

Holtedahl, H. 1965. Recent turbidities in the Hardanger Fjord, Norway. In *Proceedings of the 17th Symposium of the Colston Research Society,* W.F. Whittard and R. Bradshaw, eds. London: Butterworths. 107–140.

ICM. 1971–1999. *Annual Environmental Reports.* Port Hardy, BC: BHP Billiton Minerals Canada Ltd.

———. 1999. *Reclamation Report.* Port Hardy, BC: Utah Mining and BHP Billiton Minerals Canada Ltd.

ICM, Environmental Department. 1980. *A Study of the Suspended Sediment Movement through Quatsino Narrows.* Port Hardy, BC: BHP Billiton Minerals Canada Ltd. 4 pp.

Johnson, R.D. 1974. *Dispersal of recent sediment and mine tailing in a shallow-silled fjord, Rupert Inlet, British Columbia.* Unpublished Ph.D. diss. UBC, Vancouver. 181 pp.

McElhanney Survey & Engineering Ltd. February 1969. Pre-mine Composite Bathymetry of Rupert Inlet. Edmonton, AB: McElhanney.

Normark, W.R., and F.H. Dickson. 1976a. Sublacustrine fan morphology on Lake Superior. *Bull. Am. Assoc. Pet. Geol.* 60:1021–1036.

———. 1976b. Man-made turbidity currents in Lake Superior. *Sedimentology* 23:815–831.

Pederson, T., D. Ellis, G. Poling, and C. Pelletier. 1993. Island Copper Mine, Canada. In *Case Studies of Submarine Tailings Disposal: Volume 1—North American Examples.* G. Poling and D. Ellis, eds. U.S. Bureau of Mines Open-File Report. Washington, DC; U.S. Department of the Interior, Bureau of Mines.

Pickard, G.L. 1963. Oceanographic features of inlets in the British Columbian coast. *J. Fish. Res. Bd. Can.* 20:1109–1144.

Poling, G.W. 1979. Environmental considerations in tailing disposal. *CIM Bull.* 72(801):144–153.

———. 1982. The characteristics of mill tailings and their behavior in marine environments. In *Marine Tailings Disposal,* D.V. Ellis, ed. Ann Arbor, MI: Ann Arbor Science. 63–84.

Rescan Environmental Services Ltd. 1999. *Hope Bay Belt Project: Boston Property Tailings Characterization.* Draft Report. San Francisco, CA: BHP Billiton World Minerals.

———. 2000. *Shoreline Rock Fill and Underwater Tailings Placement from the Island Copper Mine—An Assessment of the Morphological Changes in the Sea Floor of Rupert and Holberg Inlets.* Report to BHP Billiton Minerals Canada Ltd., Island Copper Mine, Port Hardy, BC. 44 pp.

Soulby, R. 1977. *Dynamics of Marine Sands—A Manual for Practical Applications.* London: Thomas Telford. 249 pp.

Stucchi, D.J. 1980. The tidal jet in Rupert/Holberg Inlet. In *Fjord Oceanography,* H.J. Freeland, D.M. Farmer, and D.D. Levings, eds. New York: Plenum Press. 491–497.

Stucchi, D.J., and D.M. Farmer. 1976. Deepwater exchange in Rupert/Holberg Inlet. *Pac. Mar. Sci. Rep.* 76(10):31 pp.

Tesaker, E. 1975. Modeling of suspension currents. In *Proceedings of Symposium on Modeling Techniques.* San Francisco: American Society of Civil Engineers. 1385–1401.

Thomson, R.E. 1981. *Oceanography of British Columbia Coast. Canadian Special Publication of Fisheries and Aquatic Sciences.* 56:291 pp.

UBC. 1991. *Final Report—Acid Mine Drainage Study of the North Dump, Island Copper Mine.* Vancouver, BC: UBC Department of Mining and Mineral Process Engineering.

Utah Construction & Mining Co. 1969. *Brief in support of its application for marine tailings disposal dated October 2, 1969, pursuant to the British Columbia Jurisdiction Pollution Control Act, 1967.*

Utah Mines Ltd. 1984. Annual Environmental Assessment Report, Island Copper Mine. Port Hardy, BC: Utah Mines Ltd.

CHAPTER 5

Geochemistry: Chemical Stabilities of Tailings Sediment

George W. Poling

MINERALOGICAL COMPOSITION OF ISLAND COPPER MINE TAILINGS SOLIDS

Tailings derived from processing the low-grade porphyry copper-molybdenum ore at the ICM were composed primarily of quartz, feldspar, and biotite-chlorite minerals. Table 3.1 showed that these three relatively inert minerals made up as much as 90% of the tailings. Magnetite (at 2%–4%) and calcite (at ~2.5%) are also essentially inert or, in the case of calcite, yield neutralization capacity to the tailings. Pyrite, making up generally 2%–4% of the tailings solids, is the major "reactive" mineral in the tailings. The presence of pyrite can give rise to acid-generating potential as a result of weathering–oxidation reactions.

Table 3.1 includes detailed mineralogical analyses of natural bottom sediments from the seafloor of Rupert Inlet for comparison to the composition of the tailing solids. The tailing solids were significantly higher in both copper and molybdenum contents; otherwise, the two solids were exceedingly similar. This was not surprising because the natural sediments would have originated from several thousand years of weathering and glacial erosion of the surrounding country rock from which the ore also originated. Most of the residual-reactive iron sulfides, copper iron sulfides, zinc sulfides, lead sulfides, arsenic sulfides, and molybdenum sulfides in the tailings would have existed as discrete crystallites that were either fully liberated or occluded within larger oxide or silicate particles. Metal release from such tailing solids and subsequent possible bioaccumulation in the marine food chain is one of the most serious potential hazards in discharging mill tailings to the sea.

Early simulation studies were conducted to assess the potential release of heavy metals to the seawater column from the tailing solids during DSTP. Continuous agitation of typical ICM tailing solids (at 17% solids by weight) in seawater for up to 37 days demonstrated that no increase in dissolved copper or iron could be detected in the sea-water. In fact, at the end of this test, when the tailing solids were allowed to settle out and left undisturbed, the dissolved copper and iron contents of the supernatant water both decreased. The supernatant metal losses were found to balance almost exactly by corresponding increased metal contents in the interstitial waters trapped within the tailing sediment (Evans et al. 1979). The tailings in this simulation acted as a sink rather than a source for these dissolved metals. A second simulation test (also reported in Evans et al. 1979) running for 6 months reported similar findings with respect to both copper and molybdenum behaviors. However, this test did indicate that manganese would be released to the supernatant.

BC Research also conducted some early tests for the acid-generating potential of the ICM tailings. These studies indicated that ICM waste rock could be amenable to dump leaching but at too high an acid consumption (which resulted from the significant quantity of carbonate rock present). These tests probably contributed to an early belief among the mine operators that natural weathering reactions would not generate acid rock drainage (ARD) at the ICM.

TAILINGS AS MARINE SEDIMENTS

In general, metals present during discharge, such as solid metal sulfide or metal oxide crystals, must convert to free "solvated" ions before they become bioavailable and potentially toxic. Admittedly this discounts the mechanism wherein an organism might ingest some tailing particulate and the strongly acidic gastrointestinal tract might make the metal ion bioavailable. Physical stresses or abrasiveness of tailing turbidity flows or clouds will probably result in avoidance by mobile organisms before ingestion, and chemical toxicity can play a significant role. The more important question, then, is What conditions or factors promote ionization or dissolution of metal sulfides or metal oxide particulates? For those ions already in solution or simply adsorbed on the solids at the time of discharge, desorption/adsorption equilibria might control eventual release to the water column or incorporation into the sediments.

Some of the parameters that influence the bioavailability of metals in a marine environment are redox potential, pH, salinity, the water's dissolved oxygen content, the mineralogy of solid particulates, particle sizes, sedimentation rates, and the quantity and quality of organic detritus incorporated with the tailing sediments. Of these, the most important is probably redox potential, although this is often interdependent on oxygen content. Some of the less important parameters will be considered first.

Particle Size

Coarse particles settle faster and can thereby become incorporated as buried sediment particles quicker than finer particles. Thus they might be subjected to an oxygenated leaching environment for shorter periods in a DSTP system. Because coarse particles also have a lower surface area per unit weight, they will also have a lower leaching rate per unit weight. Although the ICM tailings were typically 50% coarser (and 50% finer) than 30 µm, the majority of the surface area was found on the 10% by weight of the particles finer than 5 µm.

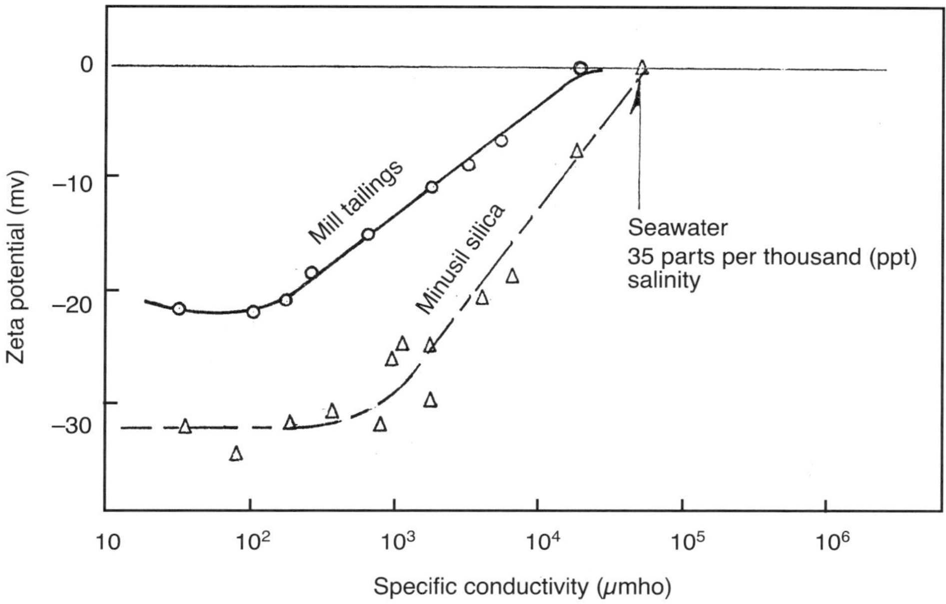

FIGURE 5.1 Effect of seawater addition on zeta potential

Salinity

Higher salinity waters, such as normal full-strength seawater, decrease the bioavailability of metal contaminants in DSTP by a variety of mechanisms. The ionic "atmospheres" surrounding each solid particle are dramatically compressed in the relatively high ionic strength (~35 ppt salinity) seawater. This so-called compression or thinning of the "electrical double layer" enables short-range Van der Waals forces (which are net attractive forces) to overcome the normally longer-range electrostatic repulsive forces of "like" charged particles, leading to coagulation of impacting particles. Coagulation promotes aggregation of particles into larger coagula (or flocs if organic detritus is also coadsorbed), which in turn settle out much faster than fine individual particles. Of particular importance is the fact that coagulation strongly promotes incorporation of the ultrafines (i.e., the -5-μm fraction with the majority surface area) within the rapidly settling aggregates or coagula. Seawater-induced coagulation thereby more rapidly incorporates both adsorbed ions and ultrafine particles within a tailing sediment. Figure 5.1 illustrates how the high salinity of seawater (increased specific conductivity) reduces the surface charges (zeta potential) of both mill tailings and reference ultrafine silica to promote coagulation. Active coagulation will normally ensue at zeta potentials of less than 10 to 15 mV, absolute (+ or −).

pH

Most metal sulfides and oxides become more soluble as the pH is reduced into acidic regimes. In turn, adsorption/coprecipitation of metal ions—Pb, Cu, Zn, and Cd—onto hydrous solids becomes much more efficient at alkaline pHs. Leckie and colleagues showed that although only a small percentage of these four metal ions are absorbed at pHs from 4.5 to 6.1, more than 90% was absorbed at pHs from 6.0 to 8.0 (Leckie et al. 1980). Seawater with a pH of 7.9 to 8.1 thus favors adsorption of most heavy metal ions and incorporation of metal contaminants within tailing sediments.

Sedimentation Rate

Higher sedimentation rates reduce the residence time of particulates within the water column, thereby reducing potential leaching times. More rapid sedimentation will incorporate a smaller proportion of organic detritus and bury it deeper within the sediment. This typically moves these sediments into reducing environments as opposed to oxidizing environments.

Mineralogies of Tailing Particulates

Metal sulfide minerals differ considerably in their rates of oxidation or conversion to ionic species. In terms of weathering reactions that can give rise to acidic drainage, minerals such as pyrrhotite, pyrite, and chalcopyrite are more reactive than galena and sphalerite. Because most sulfides are either semiconductors or electronic conductors and exhibit different surface or "mixed" potentials when in intimate contact, galvanic driving forces can accelerate the dissolution of the less noble sulfide. This is often a more important factor in release of metal ions in the instance of DSTP of massive sulfide tailings. In the ICM situation the residual metal sulfide minerals are seldom in galvanic contact. In fact, minute sulfide particles were often totally or mainly occluded within insulating and inert oxide or silicate matrices. Because of this, surface exposures of sulfides were reduced and potential leaching rates were lessened.

Incorporation of Organic Detritus Within the Sediments

Organic matter and its degradation by bacterial respiratory processes influence redox conditions within bottom sediments. It turns out that detritus derived from planktonic sources is much more readily degraded than organics of terrestrial origin. In the bottom regions where tailings sedimentation rates were high in Rupert/Holberg Inlets during the operating life of the mine (many centimeters per year), the amount of organic incorporated was very small. After mine closure, however, organic detritus would again dominate on stable sediment deposits.

Redox Potentials

Most metals and metal compounds such as sulfide minerals tend to become thermodynamically unstable under oxidizing conditions and remain stable as sulfides under reducing conditions. Most metals, then, tend to become more soluble under oxidizing conditions. For this reason, redox potential is a primary factor in determining the propensity of most metal sulfides to ionize or dissolve.

In subaqueous systems oxygen is usually the most efficient oxidizing agent available, even though it is available in much lower concentrations than in the atmosphere (air). Even in nearly saturated water, the concentration of oxygen is only 1/30 of that in air at sea level. Couple this with the low rate of oxygen diffusion in water and it is easy to

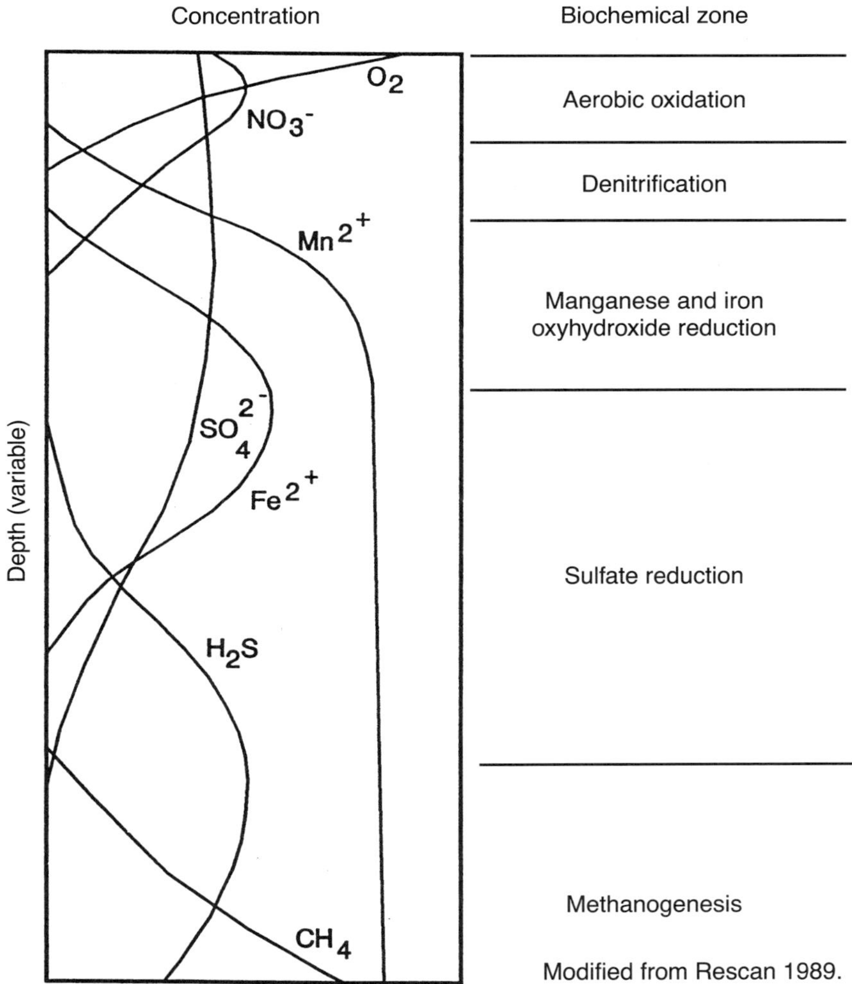

Schematic distribution of biogeochemically important molecules and ions in interstitial waters in sediments showing the zonation typically observed in marine and lacustrine sediments. This distribution will stretch or compress in response to changing physical, chemical, and biological conditions.

FIGURE 5.2   Schematic distribution of biogeochemically important species in interstitial waters in sediments (Pedersen 1985)

understand why many sea inlet bottom and particularly sea bottom sediments are anoxic and reducing in character.

When oxygen is unavailable as an oxidant, other oxidants can come into play. In bottom sediments, both chemical and biological components will progressively utilize oxygen first, nitrate second, hydrous Mn and Fe oxides third, and sulfate fourth. Figure 5.2 shows a typical zonation within the surface region of marine sediments resulting from the types of "early diagenetic" reactions mentioned previously (Pedersen 1984, 1985). Bacterial populations that decompose organic detritus normally deplete the sediments of oxygen within several millimeters of the sediment/seawater interface even when the water column is well oxygenated to the inlet bottom as it was at ICM. As one moves deeper into the sediment and $SO_4^=$ becomes the oxidant, metals will be precipitated as insoluble sulfide precipitates. Thus although the topmost region of the sediment might release metals to the water column via diffusion down the concentration gradient, most metals will be stable as sulfides within the deeper sediment layers.

Freshly forming Mn and Fe oxides near the top of the sediments adsorb and scavenge metal ions that diffuse upward and prevent their release. As the Mn and Fe oxides are buried deeper, their adsorbed (scavenged) heavy metals are again released as ionic species and in turn are reduced to sulfides (they become authigenic sulfides). This near-surface cycling of heavy metals within the surface regions is believed to effectively "cap" the sediments and help to prevent metal release.

Cores representing three different sedimentary facies were taken from Rupert and Holberg Inlets in 1980 (Pedersen 1985)—natural sediments in upper Holberg Inlet, rapidly accumulating tailing slightly down inlet from the outfall, and slowly accumulating tailings near the head of Rupert Inlet.

Distributions of dissolved copper in surface sediments were similar at all three sites. Dissolved copper concentration decreased with depth at all three sites with less decrease in the tailings than in the natural sediments. These studies indicated clearly that, at least in 1980, the tailings in Rupert Inlet were not supporting a net benthic flux of copper into the inlet waters.

Dissolved molybdenum in tailing pore waters was enriched in the rapidly accumulating deposits by a factor of 6 over normal seawater. In the slowly accumulating tailings, Mo was enriched by a factor of 2. The observed profiles indicated that the tailings were releasing some Mo to the overlying waters. Pedersen (1985) calculated that the efflux of Mo was extremely small (<0.002% of contained Mo, in the inlet). This small efflux was deemed to be insignificant when compared to the naturally relatively high concentration of Mo in seawater (~11 ppb). Thus the slight amount of molybdenum release was concluded to have no measurable effect on the waters of Rupert Inlet.

The flux of arsenic from tailings in Rupert/Holberg Inlet was very small. The finding (Pedersen 1985) of $SO_4^=$ reduction close to the sediment surface of tailing-rich sediments probably precipitated most as authigenic sulfides.

BIOASSAYS ON TAILINGS

Throughout most of the life of the mine, monthly grab samples were submitted (undiluted by seawater) to BC Research in Vancouver for acute toxicity testing using acclimated juvenile coho salmon. The salmon were exposed to a 1:1 slurry to seawater mixture for a 96-hour period. Throughout the mine's operating life these static bioassays reported 100% survival and therefore no acute toxicity except for one sample. This one sample, out of the many hundreds, reported an 80% survival rate. This deviation from 100% survival was traced to a plant test of a new frother reagent that contained some unexpected ammonia content. The ammonia solubilized copper to generate some slight toxicity in the bioassay tests. Although this one unexpected result did provoke concern among the mine operators and reinforced the need for complete bioassay testing of any potential new mill reagents, it did not create toxicity in the inlet system. This one experience, well into the operating life of this mine, does underline the need for special vigilance when using a DSTP system.

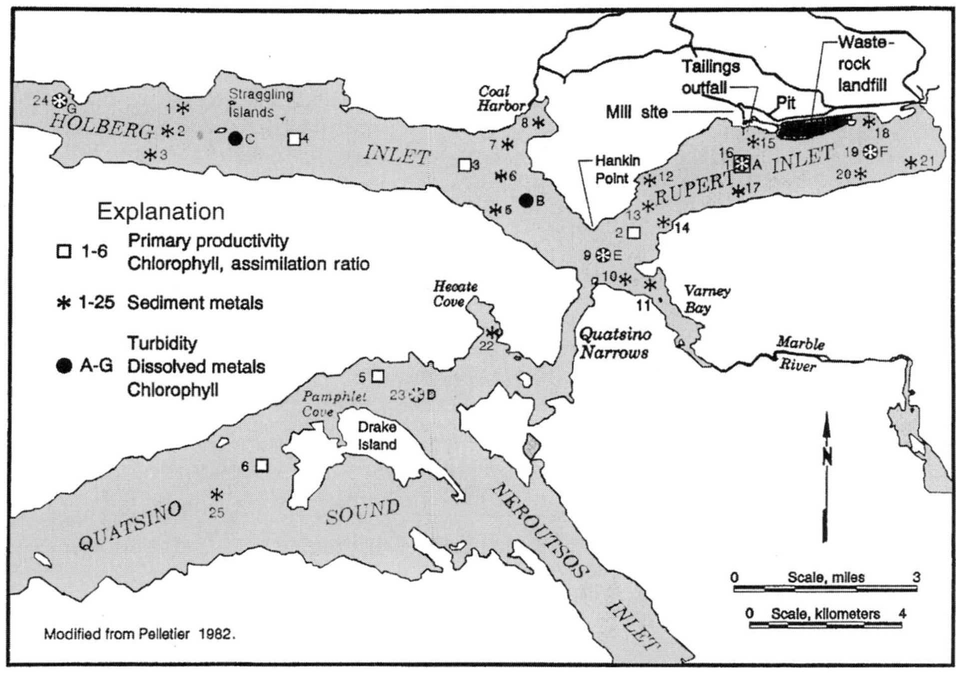

FIGURE 5.3   Rupert Inlet and adjacent fjords with selected monitoring stations

SEAWATER COLUMN CHEMICAL QUALITY

Through most of the 25-year operating life of the ICM, seven seawater sampling stations, labeled A to G on Figure 5.3, were analyzed each month to profile temperature, turbidity, color, transparency, and suspended solids. Each quarter, the seven stations were sampled to profile salinity; alkalinity; pH; dissolved oxygen; spent sulfite; total As, CN, and Hg; and dissolved and particulate Cd, Co, Cr, Cu, Fe, Mo, Mn, Ni, Pb, and Zn. Each year, the clarity of water at 120 stations was determined using a transmissometer.

Mean annual concentrations of copper showed that the ambient levels fluctuated around predischarge levels of 2 μg/L from 1971 to 2000. Figure 5.4 shows graphs for stations A, E, and F from 1971 to 1993. Most other metals behaved similarly. Manganese fluctuated from predischarge levels of ~3 μg/L to about 10 μg/L, as shown in Figure 5.5.

REFERENCES

Evans, J.B., D.V. Ellis, J. Leja, G.W. Poling, and C.A. Pelletier. 1979. Monitoring of porphyry copper tailing discharged into a marine environment. *Proceedings XIII Int. Min. Proc. Congress*, Warsaw, Poland. 650–690.

Leckie, J.O., M. Benjamin, K. Hayes, G. Kaufman, and S. Altmann. 1980. *Adsorption/Coprecipitation of Trace Elements from Water with Iron Oxyhydroxide.* Final report, EPRI RP-910. Palo Alto, CA: Electric Power Research Institute (EPRI).

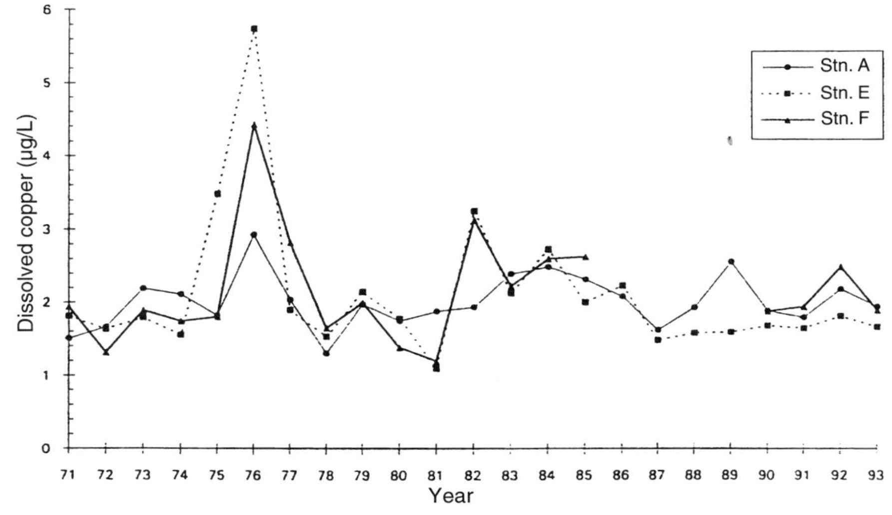

FIGURE 5.4 Dissolved copper in seawater (stations A, E, and F; 1971–1993)

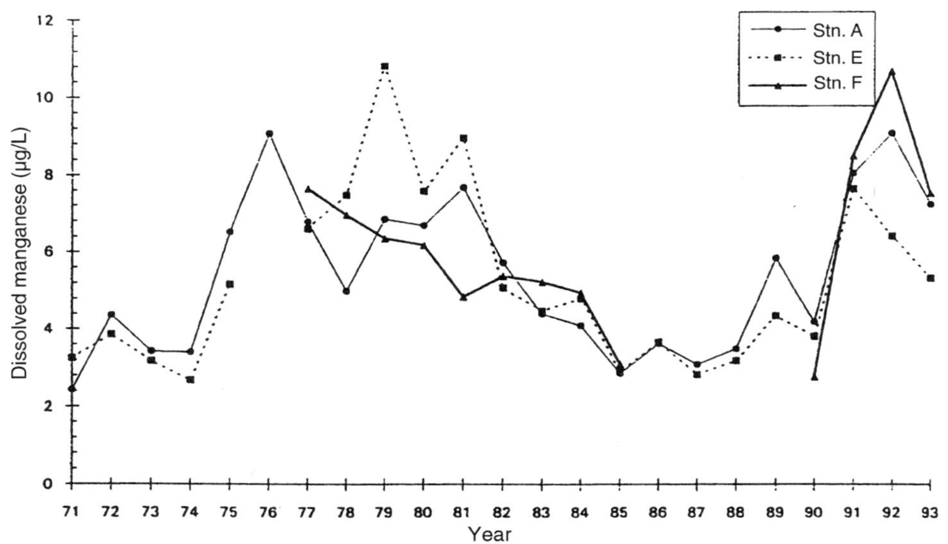

FIGURE 5.5 Dissolved manganese in seawater (stations A, E, and F; 1971–1993)

Pedersen, T.F. 1984. Interstitial water metabolite chemistry in a marine mine tailings deposit, Rupert Inlet, British Columbia. *Canadian Journal Earth Sciences* 21:1–9.

———. 1985. Early diagenesis of copper and molybdenum in mine tailings and natural sediments in Rupert and Holberg Inlets, British Columbia. *Canadian Journal Earth Sciences* 22:1474–1484.

Changes in Physical and Chemical Properties of Rupert Inlet Waters

T.R. Parsons

INTRODUCTION

Physical and chemical oceanographic properties of Rupert Inlet, Holberg Inlet, and Quatsino Sound were monitored before, during, and after the operation of the ICM in Rupert Inlet. Data gathered 1 year before operation have been compared with data collected during 22 years of mining (1971–1992; ICM 1971–1998) and with data obtained after the mine closed in 1995. The 22 years of data monitored before mine closure are reported and analyzed in two reports (Zeng and Parsons 1994a, 1994b); more recent post-closure data referred to in this chapter are taken from the ICM annual reports starting in 1996 (ICM 1971–1998).

All data were collected to determine if there were any long-term changes in oceanographic properties that could be attributed to mining activity. The frequency of data collection (usually 4 to 12 times a year) did not lend itself to any dynamic interpretation of the data. Consequently, the analysis that has been performed is purely a statistical comparison of data collected at different stations close to the mine, and at some distance from the mine in Holberg Inlet and Quatsino Sound. Data gathered cover the properties of the seawater (temperature, salinity, and turbidity) and dissolved metals that could possibly be attributed to mining activity.

Changes in physical and chemical parameters in Rupert Inlet and the surrounding pelagic system did not indicate that the ICM had any large-scale or long-term effect on the waters of this region. Although some turbidity values could be attributed to mining, a shallowing of the euphotic zone did not occur. Temperature and salinity data did not indicate any change in water circulation in the area. Dissolved copper showed no long-term change in concentration. Zinc and manganese showed opposite trends, which could have been attributable either to mining or other anthropogenic activities in the inlet system.

METHODS

Field Sampling

Seawater samples were collected 12 times a year at six stations (Figure 6.1) using 5-L polyvinyl chloride (PVC) Niskin samplers. Samples were taken at 0, 5, and 30 m, as well as at the bottom depth at each station. Temperature was taken at depth with a SIS RTM4002 digital thermometer attached to the Niskin sampler. Seawater samples were analyzed for salinity by titration (Strickland and Parsons 1972); turbidity was measured nephelometrically (APHA, AWWA, and WPCF 1985) and with a Secchi disc; metals were measured using flame A.A.S. Protocols for these analyses are given in the ICM *Methods Manual* (1986). Commercially available metal standards (certified by the National Research Council) were run with all metal samples for quality control.

Statistical Analyses

Conventional statistical tests were performed on the data, including analysis of variance (ANOVA), correlation analysis, and linear regression. The probability test for ANOVA has been based on a significant difference between means that are greater than 95%. Correlations and regressions were also based on a significance of 95%. The ANOVA test was used to compare variable means in the data collected from different stations, the correlations were used to look for possible relationships between

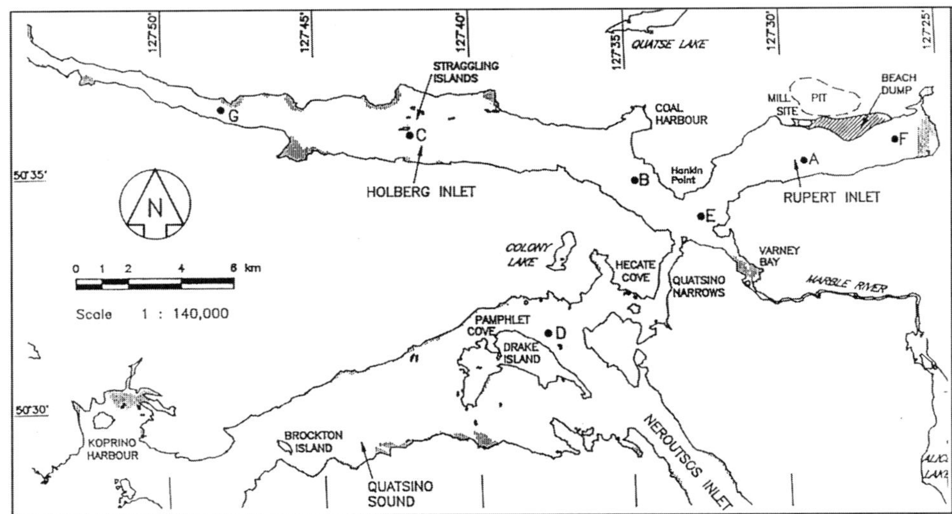

FIGURE 6.1   Station locations

variables, and the linear regressions of time series data were employed to indicate any long-term trend in a property.

RESULTS AND DISCUSSION

Comparison of Means Before and After Starting Mining Operations

Zeng and Parsons (1994a) compared average results for temperature, salinity, transparency, and metal concentrations from data collected before and after mining started. Unfortunately the number of data points collected before mining was less than 10 for each parameter, but there were generally between 100 and 250 data points collected after mining started. Because the interannual differences in the above parameters are large, the comparison by Zeng and Parsons (1994a) does not give a clear picture of the impact of mining. However, in general there were no changes in temperature, salinity, copper, or arsenic in the waters of the inlet system after mining began. Transparency showed an increase in the Secchi disc value at station A; turbidity increased at some stations (including station A) and decreased at others. Manganese concentration showed a small (<2 ppb) increase at two stations, and zinc showed a decrease at five stations. The possible significance of these results is discussed below in terms of the long-term (22-year) data trends.

COMPARISON OF DATA DURING MINE OPERATIONS

Temperature and Salinity

Temperature and salinity data were collected throughout the Rupert/Holberg/Quatsino system to determine the characteristics of water movement in the area. Average temperatures and salinities fluctuated with offshore conditions and were generally warmer and less saline within the inlets compared with Quatsino Sound. A significant increase in seawater temperature (by about 1°C) was found over the whole area during the study period. A small increase in salinity was generally correlated with the temperature rise (Zeng and Parsons 1994a). The largest changes in both these properties were seasonal, with a mean temperature range of approximately 8° to 12°C and a salinity change of approximately 27 to 32. As these changes are consistent with seasonal changes on the west coast of Vancouver Island, no fluctuation in these properties could be attributed to mining activity.

Mining activity does not seem to have affected the circulation of Rupert and Holberg Inlets even though the bottom of Rupert Inlet became shallower because of the deposition of mine tailings. Calculations on the effect of narrowing Rupert Inlet at the rock dump site (near station A) were made to see if this had caused any significant change in currents in this area. Calculations revealed that a frontal zone (Pingree, Holligan, and Mardell 1978) had not formed in this area.

Secchi Disc and Near-Surface Turbidity

As shown in Figure 6.2, both the Secchi depth and turbidity varied seasonally. Both were also strongly correlated with temperature and salinity (Zeng and Parsons 1994a). This indicates that both measures of transparency were largely governed by natural events.

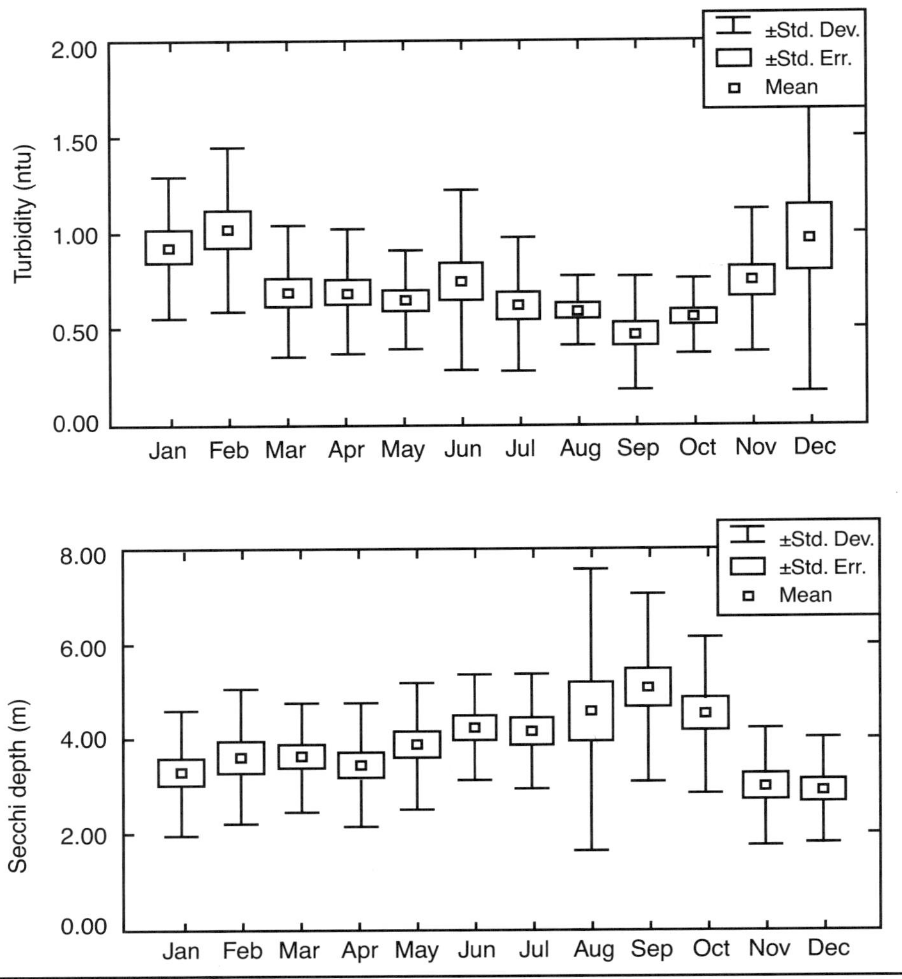

FIGURE 6.2 Average turbidity and Secchi disc depth at station A, 1971–1992, during each month of the year

Although statistical differences among stations could be detected (Zeng and Parsons 1994a), no consistent difference could be attributed directly to mining. In Rupert Inlet, station F (Figure 6.1) had the highest average turbidity and the lowest Secchi depth, and it was located closest to an estuary. Stations A and E had the next highest average turbidities and lowest Secchi depths. This could have resulted either from downstream effects of the estuarine conditions at station F or from turbidity caused by the mining activities. Turbidity from the mine could come from either sporadic upwelling at station E or from the rock dump site near station A. At both locations, turbidity could be seen on different occasions. Because there was no long-term increase in near-surface turbidity in Rupert Inlet (Table 6.1), it is probable that because the rock dump site was a more or less constant presence, it was this site that added to the estuarine turbidity in Rupert Inlet instead of the mine tailings (the tailings increased over time). However, deep water turbidity (>30 m) was always greater in Rupert Inlet, and the occasional upwelling of this turbidity at Hankin Point would account for the sporadic visual appearance of tailings at that location.

TABLE 6.1 Results from the linear regression of variables over time for data from 1971 to 1992

Variable	Station	N	A	B	R	F	p
T	A	246	9.13	0.000151	0.2267	13.324	0.00032
	B	250	9.17	0.000149	0.2269	13.459	0.00039
	C	248	9.07	0.000160	0.2570	17.406	0.00004
	D	249	8.99	0.000162	0.2497	16.426	0.00007
	E	246	9.01	0.000152	0.2382	14.674	0.00016
	F	246	9.10	0.000172	0.2466	15.804	0.00009
	G	220	9.13	0.000121	0.1881	7.999	0.00512
S	A	236	28.62	0.000100	0.1016	2.443	0.11940
	B	236	28.48	0.000130	0.1470	5.169	0.02390
	C	235	28.48	0.000120	0.1233	3.601	0.05897
	D	233	29.70	0.000050	0.0762	1.349	0.24668
	E	232	28.72	0.000080	0.0738	1.261	0.26270
	F	232	28.65	0.000080	0.0749	1.297	0.25594
	G	216	28.00	0.000190	0.1505	4.959	0.02700
SDD	A	247	4.10	−0.000041	0.0573	0.807	0.37004
	B	250	4.85	0.000016	0.0169	0.072	0.78930
	C	247	4.97	0.000038	0.0343	0.290	0.59088
	D	249	4.72	0.000004	0.0053	0.007	0.93411
	E	245	4.09	0.000037	0.0667	1.078	0.30012
	F	246	4.05	−0.000039	0.0502	0.618	0.43264
	G	218	4.69	0.000035	0.0264	0.151	0.69844
TB	A	252	0.73	−0.000002	0.0116	0.034	0.85462
	B	251	0.57	0.000002	0.0103	0.026	0.87149
	C	251	0.38	0.000004	0.0349	0.305	0.58125
	D	250	0.49	0.000028	0.1768	8.003	0.00505
	E	248	0.66	0.000018	0.1138	3.231	0.07361
	F	248	0.70	0.000008	0.0371	0.339	0.56112
	G	222	0.50	−0.000015	0.0653	0.943	0.33265

(Table continues on next page.)

TABLE 6.1   Results from the linear regression of variables over time for data from 1971 to 1992 (continued)

Variable	Station	N	A	B	R	F	p
Cu	A	88	1.75	0.000044	0.1714	2.603	0.11031
	B	79	1.66	−0.000002	0.0071	0.004	0.96076
	C	79	1.67	−0.000007	0.0334	0.086	0.77035
	D	87	1.29	0.000033	0.1379	1.648	0.20279
	E	54	1.56	0.000028	0.0998	0.523	0.47267
	F	37	1.65	0.000039	0.1786	0.153	0.29023
	G	14	1.46	0.000040	0.1264	0.195	0.66680
Mn	A	87	4.06	0.000303	0.2138	4.070	<u>0.04680</u>
	B	78	3.96	0.000096	0.1189	1.090	0.29971
	C	78	3.59	0.000225	0.2265	0.111	<u>0.04610</u>
	D	87	3.21	0.000135	0.1655	2.395	0.12544
	E	53	3.08	0.000258	0.3829	8.762	<u>0.00466</u>
	F	36	2.78	0.006260	0.6421	23.852	<u>0.00002</u>
	G	14	6.62	−0.000356	0.1887	0.443	0.51817
Zn	A	81	3.75	−0.000135	0.1633	2.165	0.14516
	B	73	5.51	−0.000446	0.3821	12.140	<u>0.00085</u>
	C	73	7.32	−0.000910	0.2935	6.693	<u>0.01173</u>
	D	81	4.16	−0.000356	0.3859	13.827	<u>0.00037</u>
	E	45	4.52	−0.000367	0.4957	14.008	<u>0.00054</u>
	F	28	4.67	0.000172	0.1424	0.538	0.46991
	G	7	4.58	0.000106	0.0373	0.007	0.93662
As	A	72	2.83	−0.000164	0.2625	5.181	<u>0.02589</u>
	B	37	2.67	−0.000076	0.1044	0.386	0.53869
	C	73	2.78	−0.000163	0.2745	5.790	<u>0.01870</u>
	D	74	1.98	−0.000039	0.0915	0.608	0.43813
	E	53	2.69	−0.000179	0.4140	10.549	<u>0.00206</u>
	F	34	2.40	−0.000144	0.3026	3.225	0.08199
	G	21	2.05	−0.000063	0.0717	0.098	0.75729

Notes:
Linear coefficients with $p < 0.05$ are underlined, indicating that those linear trends are significant. Here N = the number of observations; A = the intercept; B = the linear coefficient; R = regression coefficient; and F = the parameter for f-test.

The depth of the Secchi disc is a measure of the euphotic zone. Figure 6.2 shows that the seasonal range of this value was from 3 to 5 m, with the deepest euphotic zones observed during August and September. During 22 years of mine operation, there was no statistically significant increase in the Secchi depth at any station (Table 6.1). Following closure (Table 6.2) there was a statistically significant deepening of the euphotic zone throughout the inlet system. However, a similar event had been observed from

TABLE 6.2 A summary of differences between means of Secchi disc depth and turbidity before and after closure

Secchi Disc	1993–1995	1996–1998
Number of samples	9	9
Mean depth (m)	4.44	5.25*
Standard deviation	0.35	0.46
Turbidity		
Number of samples	9	9
Mean turbidity (ntu)[†]	0.89	0.56*
Standard deviation	0.26	0.11

* Means are significantly different at P = 0.05.

[†] ntu: nephelometric turbidity unit.

1984 to 1986 when mean Secchi depths of greater than 7 m were observed. We can conclude, therefore, that the euphotic depth of the inlet system was not affected by the operaton of the mine.

Turbidity is a measure of finely suspended particulate matter. Figure 6.2 shows that turbidity was highest (>1 ntu) during the winter (November to February) and lowest (approximately 0.5 ntu) during the late summer. At one station (Table 6.1, station D), turbidity showed a statistically significant increase throughout the period of mining. This station, located in Quatsino Sound, generally had the clearest ocean water. The turbidity increase appears to be directly attributable to the mine, resulting from some upwelling and export of tailings at station E. Post-closure turbidity values (Table 6.2) were significantly lower, including station D where values decreased from >0.8 ntu in 1994 to <0.5 ntu in 1998. In conclusion, the mine did cause definite long-term increases in turbidity in Quatsino Sound. Sporadic visual increases in turbidity could be attributed to biological activity (plankton), the rock dump site (station A), occasional upwelling at station E, and silt from the river near station F.

Changes in Metals During Mining Operations

Table 6.3 shows the mean values of copper, manganese, zinc, and arsenic found at the six stations over the period from 1971 to 1992. Typical variations in these levels for station A are shown in Figure 6.3. Linear regressions of these metals with time over 22 years are shown in Table 6.1 and Figure 6.2. In general, dissolved copper values fluctuated in a narrow range from 1–3 µg/L showing a small increase over 22 years that was not statistically significant (p = 0.05). However, Table 6.3 shows that the small difference (0.2 µg/L) in the mean copper concentration at station A was statistically significant compared to all other stations except for station F. This may have been due to a small amount of contamination from the dock loading facility where copper concentrate was loaded for shipping or it could have come from the rock dump along the shore of the inlet.

A statistical analysis of the data in Table 6.3 shows that at station A, Cu was significantly higher than at all other stations except F; at station A, Mn was significantly higher than at all other stations except F; at station F, Zn was significantly higher than at all other stations, and no clear significant difference between stations was seen for As (Zeng and Parsons 1994a).

TABLE 6.3   A summary of mean metal concentrations (µg/L) during the period from 1971 to 1992

	Metals							
	Cu		Mn		Zn		As	
Station	Mean	Standard Deviation	Mean	Standard Deviation	Mean	Standard Deviation	Mean	Standard Deviation
A	1.9	0.6	5.5	3.3	3.1	1.9	2.2	1.4
B	1.7	0.5	4.4	1.9	3.1	2.7	2.5	1.8
C	1.7	0.5	4.7	2.4	2.6	1.7	2.2	1.3
D	1.4	0.6	3.8	1.9	2.4	1.9	1.9	0.9
E	1.7	0.8	4.3	1.9	2.6	1.8	2.0	1.1
F	1.8	0.7	5.2	3.1	5.1	4.2	2.1	1.1

Dissolved manganese increased statistically at stations A, C, E, and F from less than 5 to 7 µg/L during the 22-year period. Seawater normally contains about 2 µg/L, but total manganese in river water can be around 10 µg/L. Although the source of manganese may have been mining (Table 6.3, station A was generally higher than all other stations), it could also have been released from the soil as a result of logging throughout the shoreline system during the same period. Dissolved zinc and arsenic showed sporadic increases (up to 12 µg/L for zinc and 7 µg/L for arsenic), but statistically, both elements decreased at most stations over 22 years (Table 6.1 and Figure 6.3). Station F (Table 6.3) had a statistically higher zinc concentration than other stations, which is unexplained. Because this station was close to a small river, the zinc could possibly have come from fresh water.

Changes in Dissolved Metals Following Mine Closure

Table 6.4 shows a comparison of mean values of copper, manganese, and zinc throughout the system, 3 years before closure (1993–1995) and for 3 years after closure (1996–1998). The levels of copper and manganese showed statistically significant decreases; zinc showed a statistically significant increase. The decrease in copper over the 6-year period is quite small and had been experienced previously during mining operations in 1978–1979 when some copper concentrations in the system dropped to below 1 ppb. This effect, therefore, was not unique to closure, and in general the post-closure value of 1–3 ppb was within the range of values encountered during mining operations. The decrease in the manganese and the increase in zinc concentrations are much larger, and they reflect trends opposite to those encountered at the onset of mining and during mining, when manganese increased and zinc decreased. The concentrations of these two elements may be attributed, then, to mining. However, at no time was there any apparent biological effect from the changes in concentration of these elements.

In addition to the regular sampling stations in the system, heavy metals were also monitored close to the rock dump along the beach front of the mine (Table 6.5). The values for copper (3.25 ppb) and zinc (8.8 ppb) 3 years before mine closure were much higher than values generally found throughout the inlet (Table 6.3). In the 3 years after closure, both copper and zinc values decreased significantly (2.01 ppb for copper and 6.0 ppb for zinc) at these locations. Manganese was also measured sporadically at the

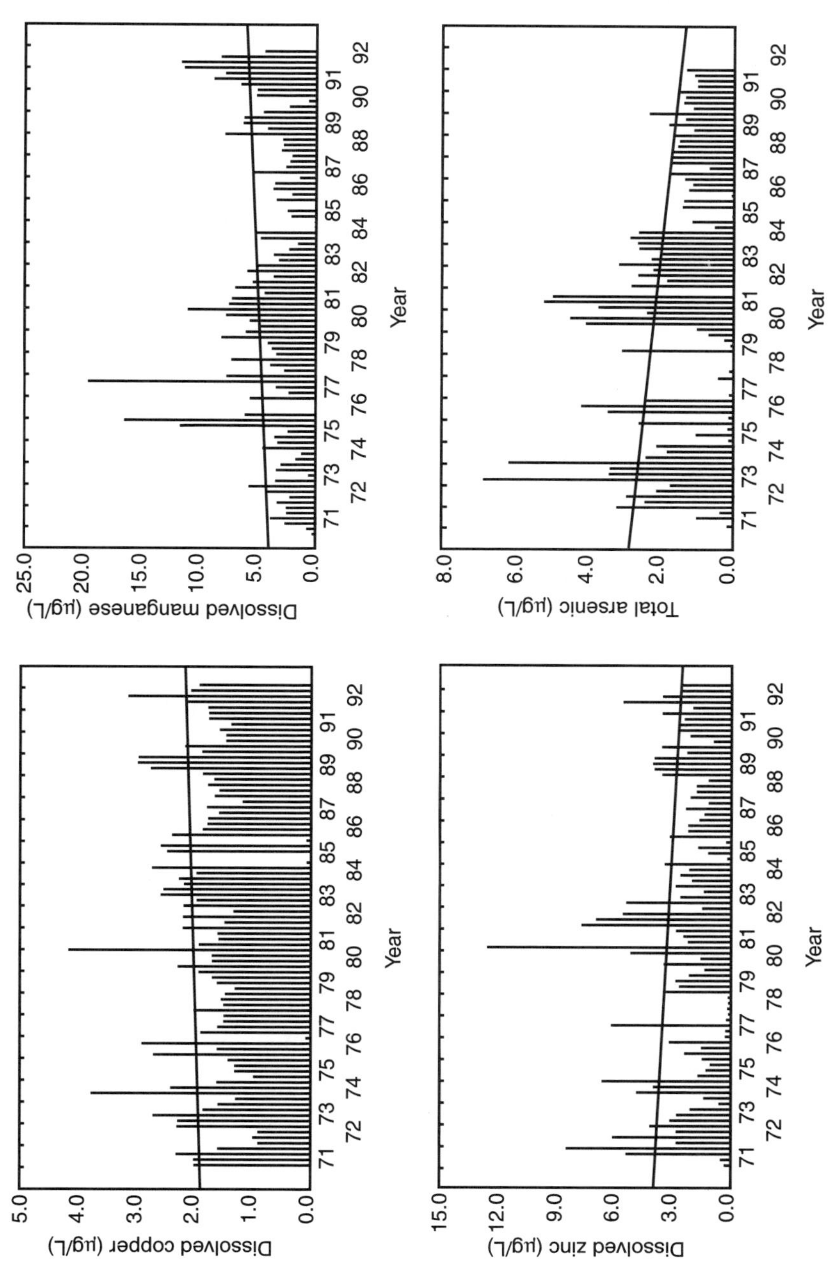

FIGURE 6.3 The bar plot with linear fit for dissolved copper, dissolved manganese, dissolved zinc, and total arsenic observed at station A from 1971–1992

TABLE 6.4 Comparison of the mean of metal levels in Rupert Inlet, Holberg Inlet, and Quatsino Sound during 1993–1995 (before closure) and during 1996–1998 (after closure)

	1993–1995	1996–1998
Copper (μg/L)		
Number of samples	18	18
Mean concentration	1.66	1.31*
Standard deviation	0.28	0.29
Manganese (μg/L)		
Number of samples	18	18
Mean concentration	5.27	2.88*
Standard deviation	1.00	0.93
Zinc (μg/L)		
Number of samples	18	18
Mean concentration	2.06	4.02*
Standard deviation	0.45	2.05

*The means are significantly different at $p = 0.05$.

TABLE 6.5 Comparison of the means of metal levels at stations close to the rock dump in Rupert Inlet, 1993–1995 (before closure) and 1996–1998 (after closure)

	1993–1995	1996–1998
Copper (μg/L)		
Number of samples	9	9
Mean concentration	3.25	2.01*
Standard deviation	0.87	0.29
Zinc (μg/L)		
Number of samples	9	9
Mean concentration	8.8	6.0*
Standard deviation	2.78	2.18

* The means are significantly different at $p = 0.05$.

beach face dump. Values in the 1980s were generally >8 ppb but were considerably lower in 1996 and 1997 with a post-closure range of approximately 4–8 ppb. This would indicate that fresh rock on the dump site released some metals and that this has declined over time, following closure. In spite of the higher values for metals immediately along the beach dump during the mining operations, the formation of a large benthic community along the rock face of the dump was not prevented.

SUMMARY

In summary, changes in temperature and salinity of the inlet system did not indicate any change in the physical environment that resulted from mining. A persistent change in turbidity in the clearest ocean water at one station over 22 years could be attributed to mining, but this did not affect the euphotic depth in the system. Increased turbidity in Rupert Inlet, which remained constant over time, may be attributed to the rock dump site.

There is some evidence that dissolved metal changes in the waters of Rupert Inlet, Holberg Inlet, and Quatsino Sound were associated with mining. The abrupt decrease in zinc after mining began, the continued decrease in both zinc and arsenic over time, and the increase in manganese over time could all have resulted from an unexplained effect of mining (especially the dumping of waste rock in the inlet). However, these changes could also have been caused by changes in freshwater runoff such as those Waldichuck (1983) documented for the Fraser River. Waldichuck showed large changes in heavy metals in estuary waters. Significant time series departures from a mean value of a data set can also be caused by oceanographic conditions, by logging activity, or by changes in the mineral composition at the rock dump site. However, whatever the cause, none of the changes reported could be considered harmful to the biological community previously as reported separately in this volume.

ACKNOWLEDGMENTS

The data in this chapter were collected by many people, some employed in the ICM Environmental Department and some employed by independent companies in the lower mainland of British Columbia. These individuals are generally recognized in the data reports cited in this chapter, and their contributions are gratefully acknowledged.

REFERENCES

APHA, AWWA, and WPCF (American Public Health Association, American Water Works Association, and Water Pollution Control Federation). 1985. *Standard Methods for Examination of Water and Wastewater.* 16th ed. Washington, DC: APHA. 1268.

ICM. 1971–1998. Annual Environmental Reports. Port Hardy, BC: ICM.

———. 1986. *Methods Manual.* Environmental Department. Port Hardy, BC: BHP Billiton Utah Mines Ltd. 106.

Pingree, R.D., P.M. Holligan, and G.T. Mardell. 1978. The effect of vertical stability on phytoplankton distributions in the summer on the northwest European shelf. *Deep-Sea Res.* 25:1011–1028.

Strickland, J.D.H., and T.R. Parsons. 1972. A practical handbook of seawater analysis. *Fish. Res. Bd. Canada Bull.* 167:310.

Waldichuck, M. 1983. Pollution in the Strait of Georgia: A review. *Can. J. Fish. Aquat. Sci.* 40:1142–1167.

Zeng, J., and T.R. Parsons. 1994a. An analysis of biological and chemical data in the near surface waters of Rupert Inlet, Holberg Inlet, and Quatsino Sound, 1971–1992. Port Hardy, BC: BHP Billiton Minerals Canada Ltd.

———. 1994b. An analysis of data on zooplankton in the waters of Rupert Inlet, Holberg Inlet, and Quatsino Sound, 1971–1992. Port Hardy, BC: BHP Billiton Minerals Canada Ltd.

Changes in the Biological Properties of the Pelagic Environment of Rupert Inlet Waters During 22 Years of Mine Operations

T.R. Parsons

INTRODUCTION

Biological oceanographic properties of Rupert Inlet, Holberg Inlet, and Quatsino Sound were monitored before, during, and after the ICM was operated in Rupert Inlet. Data gathered 1 year before mining began have been compared with data collected during 22 years of mining (ICM 1971–1992) and with 3 years of data obtained after the mine closed in 1995. These data are taken from the Annual Environmental Report Series published by ICM. The 22 years of data monitored before the mine closed are reported and analyzed in two data reports (Zeng and Parsons 1994a, 1994b). More recent postclosure data referred to in this chapter are taken from the Annual Environmental Reports (ICM 1996–1998).

All data were collected to determine if there were any long-term changes in oceanographic properties that could be attributed to mining activity. The frequency of data collections (usually 4 to 12 per year depending on the property) did not lend itself to any dynamic interpretations. Consequently, the analyses that have been performed here and in the reports cited previously are purely a statistical comparison of data collected at different stations close to the mine, as well as at some distance from the mine in Holberg Inlet and Quatsino Sound. This chapter deals with changes in biological properties, such as dissolved oxygen, chlorophyll, and zooplankton abundance.

No changes in the plankton community could be attributed to mining. In general, data on phytoplankton, zooplankton, oxygen, pH, and alkalinity were similar to those found in other communities on the coast of British Columbia. No relationships were found between the plankton populations and changes in heavy metals or turbidity in the system. Some eutrophication occurred in the inlet system during the 22 years of study. This might be attributed to increased anthropogenic activity in the area, such as fish farms, logging, housing, boating, and mining.

METHODS

Field Sampling

Seawater samples were collected 12 times a year at seven stations (Figure 7.1) using 5-L PVC Niskin samplers. Samples were taken at 0, 5, and 30 m, as well as from the bottom at each station. Alkalinity, pH, chlorophyll A, and oxygen were measured as described in Parsons, Maita, and Lalli (1984).

Zooplankton samples were usually collected in March, June, September, and December using a plankton net of 50-cm diameter and 153-μ mesh hauled vertically from the bottom to the surface. Samples were collected at four stations—A, B, C, and D (Figure 7.1). Zooplankton samples were preserved in 5% formalin and analyzed for numbers and species; diversity was calculated using Margalef's (1951) equation, $d = (S-1)/\ln N$, where S stands for the number of species in a population and N is the total count of individuals.

Statistical Analyses

Conventional statistical tests were performed on the data, including ANOVA, correlation analysis, and linear regression. The probability test for ANOVA has been based on a significant difference between means being greater than 95%. Similarly correlations and regressions were based on a significance of 95%. The ANOVA test was used to compare variable means in the data; the correlations were used to check on what factors were affecting the amount of chlorophyll A or zooplankton; the linear regressions of time-series data provided an indication of any long-term trend in a property. Any such trend, whether positive or negative, can be interpreted as a true effect or as the possible result of small changes in analytical or sampling frequency over the study period.

RESULTS AND DISCUSSION

Chlorophyll A, pH, Alkalinity, and Oxygen

Chlorophyll A can be taken as a measure of phytoplankton abundance, and changes in pH and alkalinity can sometimes be used to reflect phytoplankton productivity. Table 7.1 shows levels of chlorophyll A at four stations before mining began in 1971 compared with values collected during the next 22 years. There is no significant change in the mean of these values. The same was true for changes in pH and alkalinity, before and after mining started.

Throughout the period of study, all three parameters showed strong seasonal changes, such as those shown for chlorophyll A in Figure 7.2 at station A. However, during the 22-year period, no difference was observed in these parameters among stations in Rupert Inlet, Holberg Inlet, or Quatsino Sound that could be attributed to mining activity (Zeng and Parsons 1994a). At all stations throughout the region (Table 7.2),

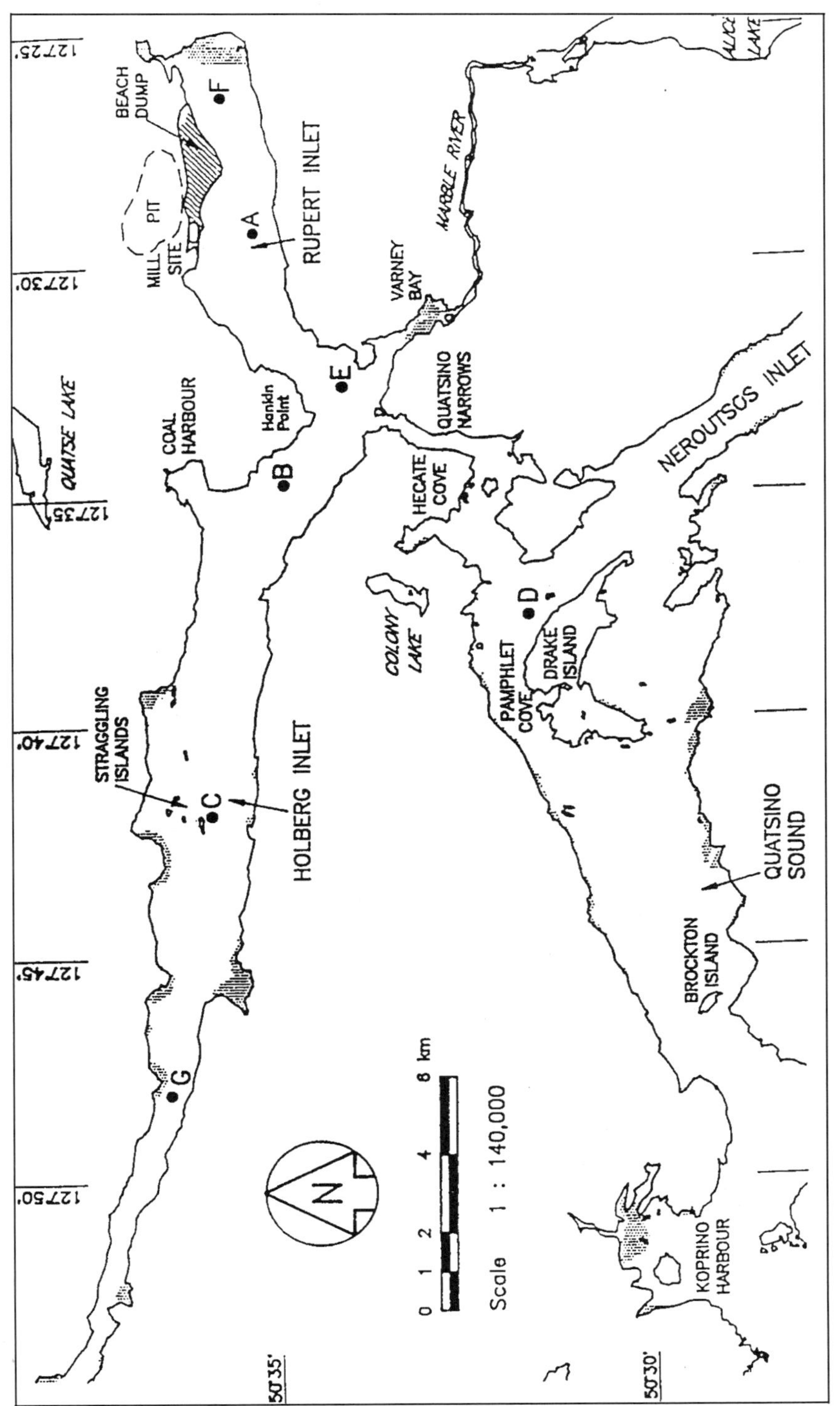

FIGURE 7.1 Station locations

TABLE 7.1 Comparison of variable means before and after September 1971 by ANOVA

		Statistics before September 1971				Statistics after September 1971			
Variable	Station	Valid N*	Mean	Standard Deviation		Valid N	Mean	Standard Deviation	p
Chlorophyll A	A	7	3.15	6.29		209	2.63	4.31	0.75
	B	7	2.90	3.08		209	2.4	4.02	0.74
	C	7	3.04	4.00		209	2.9	5.04	0.75
	D	7	1.92	1.67		208	1.56	2.33	0.68

* N = the total count of individuals (from Margalef 1951).

TABLE 7.2 Correlation between variables from data obtained over the years from 1971 to 1992

Var.	Stn.	Chl. A	T*	S†	pH	SDD‡	TB§	ALK**	DO††	Cu	Mn	Zn	As
Chl. A	A	1.00	0.30	0.11	0.35	−0.08	0.08	0.12	0.52	−0.01	−0.17	0.05	0.08
	B	1.00	0.31	0.14	0.28	−0.10	0.11	0.12	0.36	0.00	−0.13	−0.29	0.25
	C	1.00	0.16	0.25	0.39	−0.12	0.29	0.27	0.29	0.14	−0.24	−0.12	0.11
	D	1.00	0.35	0.21	0.31	−0.08	−0.04	0.20	0.32	0.06	−0.18	−0.09	0.13
	E	1.00	0.27	0.22	0.30	−0.01	0.04	0.15	0.34	−0.02	−0.34	0.01	0.13
	F	1.00	0.28	0.10	0.38	−0.01	0.09	0.05	0.46	−0.08	−0.27	−0.25	0.09
	G	1.00	0.15	0.22	0.44	−0.14	0.12	0.24	0.39	0.07	−0.41	−0.19	0.01

* T: temperature.
† S: salinity.
‡ SDD: Secchi disc.
§ TB: turbidity in ntu.
** ALK: alkalinity.
†† DO: dissolved oxygen.
Note: A p <0.05 (underlined) means that the probability for the two means being equal is less than 5% or that the two means are significantly different from each other. Correlations represented by those coefficients are significant.

chlorophyll A was correlated over the period of mine operations with temperature, pH, and dissolved oxygen, and generally with alkalinity and salinity. These were all correlations to be expected from the flux of nutrients and photosynthetic activity. Virtually no correlations (Table 7.2) were obtained between chlorophyll and heavy metal concentrations, which indicated that any release of metals was not affecting the growth of phytoplankton. Over a period of 22 years there was a statistical increase in the amount of chlorophyll A at all stations in the inlet system (Zeng and Parsons 1994a). This increase is consistent with industrial eutrophication in the area as a result of a variety of activities including housing development, logging, boating, and aquaculture, all of which tend to add nutrients to coastal systems (e.g., see changes in chlorophyll A in Figure 7.2 for station A). It is also possible that these changes were changes in sampling frequency. However, in either case there was no statistically significant change in chlorophyll A that could be attributed to changes in heavy metals, as shown in Table 7.2.

Oxygen values showed large seasonal differences at all stations (approximately 6–9 mg/L), and there was a seasonal correlation between oxygen and chlorophyll A values

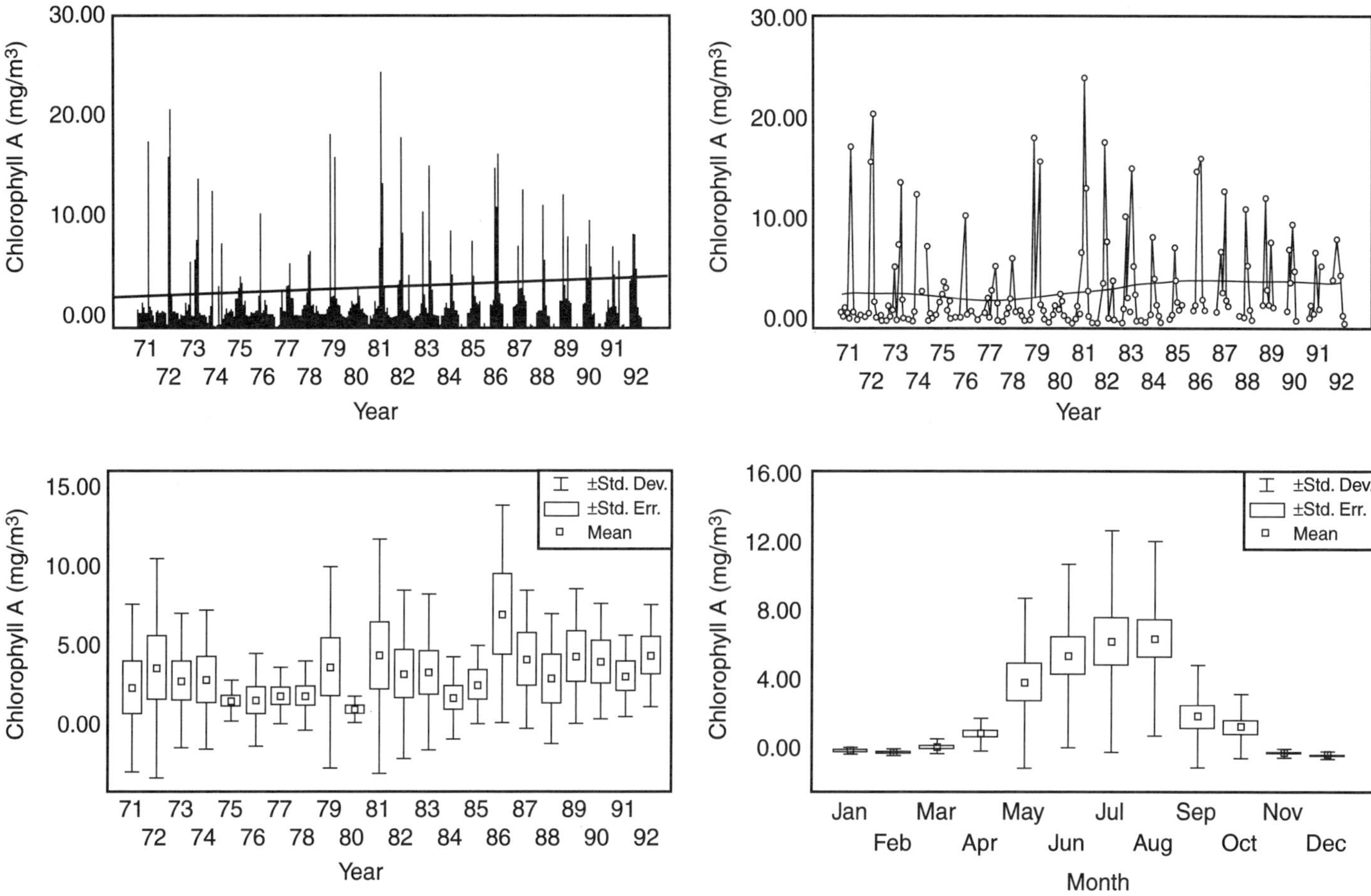

FIGURE 7.2 The concentration of chlorophyll A observed at station A from 1971 to 1992—a bar plot with linear fit (upper left), a line plot with polynomial fit (upper right), a box-whisker plot for annual means (lower left), and a box-whisker plot for seasonal means (lower right)

(Table 7.2). No statistical difference in oxygen values was seen among stations during 22 years of mine operation (Zeng and Parsons 1994a).

Total Zooplankton

Table 7.3 shows the means and standard deviations for total counts (number per tow) of zooplankton from 1971 to 1993. A statistical comparison of the means for four periods of the year at four stations can be found in Table 7.4. Because of the very patchy nature of zooplankton distributions, Table 7.4 does show some statistical differences between stations in different months, but these are not consistent with any systematic change in station A (closest to the mine site) compared with stations B, C, and D (further from the mine site).

Table 7.5 shows correlations between total zooplankton, chlorophyll A, temperature, and four dissolved metals. None of these results exhibit any consistent correlation between zooplankton abundance and other properties in the area, including the presence of heavy metals.

Table 7.6 presents linear regression results for the total counts of zooplankton at all stations for four months of the year. The results do not show any trend in abundance over time that could be attributed to mining activity. Table 7.7 shows the results of linear regression of zooplankton diversity with time. These results indicate a statistical increase in diversity at nearly all stations with time. This result is probably best explained as an analytical artifact caused by increased awareness of different species over time; in any event, there is no indication that stations either near or far from the mine site changed species diversity any differently.

TABLE 7.3 Descriptive statistics for the total counts of zooplankton from vertical hauls between 1971 and 1993

Month	Station	Mean	Std. Dev.	Minimum	Maximum	Valid *N*
March	A	30,036	62,600	1,026	295,065	21
	B	34,545	52,556	2,690	251,068	21
	C	27,700	34,160	113	165,816	21
	D	21,188	15,355	1,610	56,580	21
June	A	79,792	57,354	4,032	200,938	15
	B	116,998	79,568	10,330	288,693	15
	C	81,657	59,900	20,809	210,628	15
	D	84,875	57,854	10,383	170,314	15
September	A	83,788	70,126	4,153	271,706	22
	B	122,749	82,274	14,349	318,223	21
	C	88,286	70,798	25,898	264,467	22
	D	90,166	54,003	6,931	231,157	21
December	A	48,517	64,607	17,842	261,546	13
	B	70,672	53,076	16,294	206,216	13
	C	41,999	30,663	539	99,080	12
	D	60,215	84,444	1,934	334,661	13

Note: A vertical haul was conducted through the entire water column by a 50-cm-diameter, 153-μm mesh net.

TABLE 7.4   Results from t-tests for the total counts of zooplankton from vertical hauls between 1971 and 1993

Month	Station	A	B	C	D
March	A	1.0000	0.3122	0.7485	0.4673
	B	0.3122	1.0000	0.1571	0.1810
	C	0.7485	0.1571	1.0000	0.2554
	D	0.4673	0.1810	0.2554	1.0000
June	A	1.0000	<u>0.0474</u>	0.9109	0.7266
	B	<u>0.0474</u>	1.0000	<u>0.0445</u>	0.0819
	C	0.9109	<u>0.0445</u>	1.0000	0.8448
	D	0.7266	0.0819	0.8448	1.0000
September	A	1.0000	<u>0.0326</u>	0.6604	0.8695
	B	<u>0.0326</u>	1.0000	0.0580	<u>0.0266</u>
	C	0.6604	0.0580	1.0000	0.9433
	D	0.8695	<u>0.0266</u>	0.9433	1.0000
December	A	1.0000	0.0835	0.2017	0.1519
	B	0.0835	1.0000	0.0924	0.4716
	C	0.2017	0.0924	1.0000	0.5061
	D	0.1519	0.4716	0.5061	1.0000

Note: Significant differences between means with $p < 0.05$ are underlined.

TABLE 7.5   Correlation of total zooplankton with chlorophyll A, temperature, dissolved copper, dissolved manganese, dissolved zinc, and total arsenic

Station	Month	Chl. A	T	Cu	Mn	Zn	As
A	March	−0.3794	−0.1229	−0.2870	−0.1281	0.1942	0.0027
	June	−0.3879	<u>0.5642</u>	−0.2759	−0.0608	−0.4666	−0.0529
	September	0.2868	<u>0.4915</u>	−0.1983	0.1135	−0.2120	−0.1320
	December	−0.2521	−0.0271	−0.1244	−0.2845	−0.2446	<u>0.7374</u>
B	March	−0.1975	−0.0904	−0.3119	0.0212	0.1723	−0.3483
	June	−0.1254	0.4548	<u>0.5813</u>	0.2086	0.3685	0.5828
	September	−0.0635	0.3423	0.1885	0.1819	0.2647	−0.1749
	December	−0.3767	−0.2084	−0.0118	0.2032	−0.5211	0.0844
C	March	−0.3008	−0.0265	−0.2864	−0.0257	−0.2093	−0.0725
	June	0.2211	0.0093	−0.5427	0.4106	−0.0058	0.4555
	September	0.0886	0.0667	−0.2712	0.0362	−0.0424	0.2832
	December	−0.1708	0.4411	0.0355	0.0436	−0.5293	−0.2719
D	March	−0.3771	0.0417	−0.0423	0.1596	−0.1735	0.1123
	June	0.1372	0.4865	0.0587	0.4381	0.2824	0.4849
	September	0.3373	0.2246	−0.2058	−0.0005	−0.0200	−0.0019
	December	−0.2267	−0.1846	−0.2762	−0.3742	−0.4035	0.0278

Notes: Significant coefficients with $p < 0.05$ are underlined. Data for all variables cover the period between 1971 and 1993. Data of zooplankton were from vertical hauls and from samples at 5 m for other variables.

TABLE 7.6   Results from linear regressions for the total counts of zooplankton from vertical hauls between 1971 and 1993

Taxa	Month	Valid N	a	b	R	F	p
Total zooplankton	March	21	126,229	−1,196.30	0.21129	0.88788	0.35788
Total zooplankton	June	15	184,392	−3,483.83	0.30194	1.30403	0.27408
Total zooplankton	September	23	270,993	−2,168.97	0.26032	1.52660	0.23027
Total zooplankton	December	13	292,617	−3,022.04	0.18774	0.40186	0.53909

Notes: Here a = the intercept, b = the linear coefficient, R = the regression coefficient, and F = the parameter for F-test.

TABLE 7.7   Results from linear regressions for the diversity index of zooplankton

Month	Station	Valid N	a	b	R	F	p
March	A	21	−1.8467	0.0547	0.5833	9.7994	<u>0.0055</u>
	B	21	−3.4240	0.0763	0.6685	15.3489	<u>0.0009</u>
	C	21	−1.2598	0.0457	0.5524	8.3434	<u>0.0094</u>
	D	21	0.4476	0.0247	0.2765	1.5725	0.2250
June	A	15	−3.6544	0.0727	0.4891	4.0875	0.0643
	B	15	−6.4143	0.1103	0.7381	15.5621	<u>0.0017</u>
	C	15	−7.3911	0.1245	0.7197	13.9695	<u>0.0025</u>
	D	15	−10.0111	0.1589	0.7994	23.0210	<u>0.0003</u>
September	A	22	−3.1480	0.0737	0.5425	8.3395	<u>0.0091</u>
	B	21	−3.1252	0.0699	0.5448	8.4406	<u>0.0088</u>
	C	22	−1.6315	0.0530	0.4302	4.5411	<u>0.0457</u>
	D	21	−2.7726	0.0705	0.5052	6.8541	<u>0.0165</u>
December	A	13	−8.4034	0.1384	0.8150	21.7616	<u>0.0007</u>
	B	13	−9.0681	0.1492	0.8499	28.6142	<u>0.0002</u>
	C	13	−6.0189	0.1064	0.8116	21.2335	<u>0.0008</u>
	D	13	−6.9254	0.1196	0.8162	21.9496	<u>0.0007</u>

Notes: Linear coefficients with p <0.05 are underlined, indicating a significant trend. Here a = the intercept, b = the linear coefficient, R = the regression coefficient, and F = the parameter for F-test.

The tissue metal content of unsorted zooplankton was measured in samples collected from 1972 to 1998. Copper levels were generally less than 10 mg/kg wet weight, and zinc was generally less than 15 mg/kg wet weight. Sporadically high values (e.g., 20 mg/kg wet weight for Cu) were measured occasionally; these may have been due to sample contamination. There was no obvious trend in the data over time, and there was no apparent difference among stations.

Changes in Biological Properties After Mine Closure

Table 7.8 shows the mean and standard deviation for chlorophyll values taken at three stations and at six depths in July and September 3 years before closure and for 3 years after closure. The results show that there was no statistical difference in the mean values before or after mining activities.

Table 7.8 also shows the results of a statistical analysis of total zooplankton for samples taken in March and September at three stations, 3 years before closure (1993 to

TABLE 7.8   Changes in chlorophyll A and zooplankton numbers before and after mining in Rupert Inlet

	Before Closure (1993–1995)	After Closure (1996–1998)
Chlorophyll A (mg/m^3)		
Number of samples	18	24
Mean value	3.37	3.90*
Standard deviation	2.30	2.71
Zooplankton (number/m^3)		
Number of samples	24	24
Mean value	1,479	1011*
Standard deviation	1,196	1209

*No significant difference at the $p = 0.05$ level.

1995 inclusive) and 3 years after closure (1996 to 1998 inclusive). The results do not show any statistical change in the zooplankton populations following closure.

SUMMARY

No changes in the plankton community could be attributed to mining activity, including subsurface tailings disposal or the dumping of rock along the shoreline of Rupert Inlet. In general, values for phytoplankton and zooplankton and associated parameters such as pH, alkalinity, and oxygen were similar to those found in a data summary for the Strait of Georgia (Harrison et al. 1983). The phytoplankton data from the whole inlet system give indications that mild eutrophication occurred over 22 years; however, this may be a result of changes in sampling frequency. Eutrophication can also be attributed to increased human activity in the area such as fish farms, logging, housing, mining, and boating. Pre- and post-closure values for chlorophyll A and for total zooplankton were not statistically different, which indicates that mine closure had no effect on the plankton community.

ACKNOWLEDGMENTS

The data in this chapter were collected by many people, some employed in the ICM Environmental Department and some employed by independent companies in the lower mainland of British Columbia. These individuals are generally recognized in the data reports cited in this chapter, and their contributions are gratefully acknowledged.

REFERENCES

Harrison, P.J., J.D. Fulton, F.R.J. Taylor, and T.R. Parsons. 1983. Review of the biological oceanography of the Strait of Georgia: Pelagic environment. *Can. J. Fish. Aquatic Sci.* 40:1064–1094.
ICM. 1971–1992. Annual Environmental Reports. Port Hardy, BC: ICM.
———. 1996–1998. Annual Environmental Reports. Port Hardy, BC: ICM.

Margalef, D.R. 1951. Diversidad de especies en les communidades naturales. *Publ. Inst. Biol. Apl.* 9:5–27.

Parsons, T.R., Y. Maita, and C.M. Lalli. 1984. *A Manual of Chemical and Biological Methods for Seawater Analysis.* Toronto: Pergamon Press.

Zeng, J., and T.R. Parsons. 1994a. An analysis of biological and chemical data in the near surface waters of Rupert Inlet, Holberg Inlet, and Quatsino Sound, 1971–1992. Port Hardy, BC: BHP Billiton Minerals Canada Ltd.

———. 1994b. An analysis of data on zooplankton in the waters of Rupert Inlet, Holberg Inlet, and Quatsino Sound, 1971–1992. Port Hardy, BC: BHP Billiton Minerals Canada Ltd.

CHAPTER 8

Seabed Biodiversity at Island Copper Mine: Impact and Recovery

Derek V. Ellis

INTRODUCTION

This chapter summarizes quantitative assessments of the biodiversity of the sediment seabed (the benthos) at ICM. The biodiversity of sediment seabeds has been monitored since 1970, before mining began, through to September 1998, almost 3 years after mine closure in December 1995. The benthos monitoring, which targeted sediment-inhabiting organisms (the infauna), was a major component of the marine environmental program. Initially, the monitoring was undertaken to determine biological losses resulting from deposition of the mine tailings discharged at depth into the sea. Eventually, the monitoring was also intended to reveal the rate and pattern of recolonization of the benthos, especially after tailings discharge ended in 1995. The 29 years of monitoring the sediment benthos (1970–1998) furnishes a unique long-term data set, not only for showing impacts and recovery at ICM, but also for predicting effects and recovery at other industrial sites where the seabed may be subjected to burial by industrial residues.

Environmental impacts observed consisted of reduction in biodiversity under rapid deposition of tailings (~20 cm/yr). This reduction was limited to approximately 10 km from the outfall. The biodiversity losses recorded rarely reached total obliteration, and generally involved drops to about 10 species capable of either burrowing up through depositing tailings or acting as primary opportunist species to quickly (within 1 year) colonize stabilized tailings beds. The tailings beds were a very variable and dynamic habitat in terms of grain size and stability. About six species, mostly polychaete worms, were consistent colonizers regardless of the grain size. During mining operations the near-field area of Rupert Inlet tended to be a finer grained habitat with less sand than was present before mining began. This was especially marked in the deepest basin between Rupert and Holberg Inlets, approximately 7 km from the outfall. The naturally fine-grained sediment habitat in the more distant part of Holberg Inlet was not substantively changed by the tailing discharge. While tailings were being discharged, much of the tailings beds could stabilize, even in Rupert Inlet. The biodiversity on such stabilized tailings deposits returned within 1–2 years to numbers of species and organisms within the range of unaffected stations, but with differences in actual species present. The three sampling stations nearest the outfall (within 2 km) generally recorded the lowest biodiversity during mining operations.

By 1997, 2 years after mine closure, all the sampling stations had recovered to a sustaining ecological succession of the primary opportunist species and later colonizers. The succession—although started throughout the inlets by similar species—appeared in 1998 to be progressing toward at least two different species associations, determined in part by grain size.

Ellis (2000), Ellis and Robertson (1999), Ellis et al. (1995), and Ellis and Hoover (1990) have included results from the benthic data set in recent published reviews.

Reviews of the benthos impact and recovery in limited distribution reports are by Burd and Ellis (1994) and Burd (1999). These and other limited distribution reports on the benthos are available at <http://gateway3.uvic.ca/archives/featured_collections/ mesc/home.html> or by sending an e-mail to dvellis@uvic.ca.

A final set of benthic samples was taken in September 2000, 5 years after mine closure. These samples are currently being analyzed to show whether the successional recovery has been sustained as predicted.

In view of the hypothesis of worldwide parallel-level bottom communities (Thorson 1957), the results of the ICM surveys are expected to be significant at other sites with submarine tailings placement (Ellis and Poling 1995).

Epifaunal surveys of benthos inhabiting hard surfaces of underwater cliffs, reefs, the algal forests of shallow water, and rocky shores were undertaken using very different sampling and analytical techniques. They are described in Chapter 9.

THE SEDIMENT BENTHOS (INFAUNA) SURVEYS

At ICM, the benthos monitored quantitatively for biodiversity estimates is the infauna (Thorson 1957; see Figure 8.1), meaning the benthos that lives on or burrows into sediment seabeds (sand and mud). The two fjords adjacent to the mine (Rupert and Holberg Inlets) have characteristic wide, flat sediment troughs (Figure 8.2; see Chapter 4) and steep, rocky sides (underwater cliffs). Accordingly, the benthos attached to the underwater cliffs (epifauna and epiflora) are only a small part of the total fjord ecosystem. The greater part of the two fjords comprises sediment habitat and its infauna. Because the tailings consist of fine-grained particles, they settle as fine-grained deposits and provide a habitat physically similar to naturally deposited fine sands and silts.

Infaunal biodiversity of sedentary small animals can be assessed quantitatively (with grabs; see Figure 8.3) more accurately and expeditiously than the often large and mobile epifauna that inhabits rocky reefs and algal forests. These require scuba divers observing under field conditions (see Chapter 9), ROVs or submersibles with restricted images, or traditional towed dredges that bounce and collect erratically (Holme and McIntyre 1984).

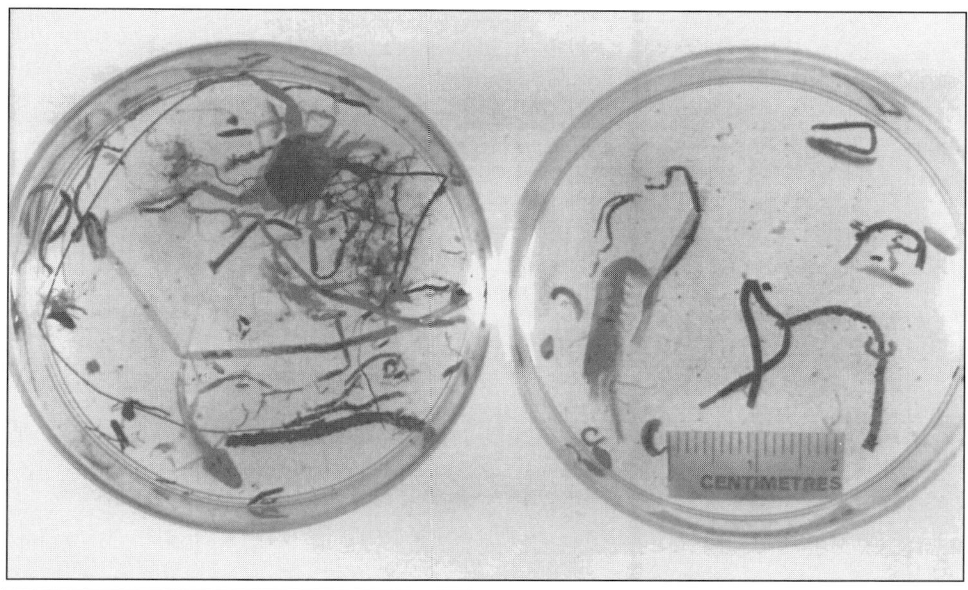

FIGURE 8.1 Example infaunal benthos: small crab, shrimp, fish, worms, and snails can be seen

To assess the infaunal biodiversity, a clamshell grab (Figure 8.3) was used to obtain samples from a constant surface area (0.05 m^2). The samples were then screened through a 30 mesh (~0.5 mm^2) sieve. From these samples, two basic biodiversity statistics were calculated (Gaston and Spicer 1998): (1) the number of species per unit area (species richness), and (2) the numbers of organisms (total number and the number of each species) per unit area (species evenness).

In 1970 before mining started, a basic sampling design of approximately 20 test and reference stations from deep and shallow water (Figure 8.2) was set. The design was to monitor (1) the targeted deposition area in Rupert Inlet into the deep basin with Holberg Inlet (generally >100 m depth; see Anon 1970), (2) shallow water sediments ringing the targeted deep area, and (3) some remote areas in Holberg Inlet and Quatsino Sound expected to be unaffected and therefore able to function as reference stations for comparisons. With minor changes and additions to eventually result in 26 sampling stations, the design was continued until September 1998, more than 2½ years after the mine closed in December 1995. The sampling design, then, embodied the monitoring principles of before and after sampling, as well as of reference and test stations (Ellis 1989).

Usually three replicates were collected at each sampling station. From 1970 to 1972 only one replicate sample at each station was sorted to species, the species were identified by scientific name, and the numbers of each species were counted. Since 1977 all replicate samples (three from each sampling station) have been processed in that way. Details of sampling procedures and statistical summaries each year are available in the annual environmental reports, especially those for 1986 and 1997 (ICM 1970–1999).

In the early years of monitoring during mining operations (1972–1976), several procedural changes were made in attempts to increase time and cost-efficiency; for example, by not identifying all specimens to species. These first attempts, in the then relatively new field of monitoring benthic responses to industrial waste discharges, were

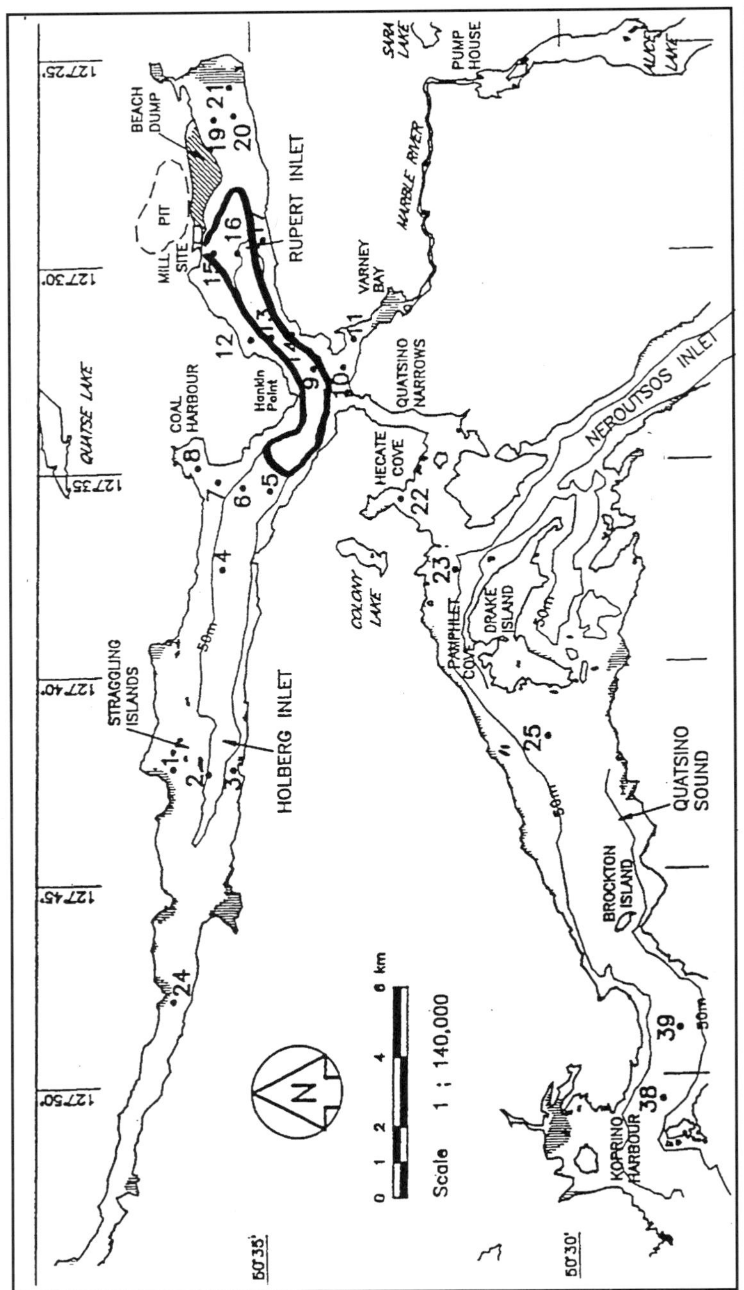

FIGURE 8.2 Station sampling positions (numbered) are located around the area targeted for tailings deposition (<100 m depth)

FIGURE 8.3   Benthic sample equipment of the type used at ICM

not satisfactory. In particular, procedures tried during the years 1973–1976 (when no collected specimens were identified to species) render the data obtained during that period generally unusable. These data are not drawn on here. Fortunately an M.Sc. thesis project carried out during 1973 and 1974 yielded some species data (Jones 1974; Jones and Ellis 1976; Ellis, Samoszynski, and Jones 1991). The early specimens also are no longer available because they were destroyed by a fire in the storage warehouse. By 1977 the definitive procedure was set: All replicate samples were sorted, identified, and counted by species. By 1996 an effective procedure in terms of time and cost had been developed (described by Ellis and Macdonald [1998] as a Rapid Preliminary Assessment Procedure). This procedure was used in 1997 and 1998.

To ensure the availability of specimens for subsequent comparisons, they have been archived in the Royal BC Museum, Victoria, Canada, and the Los Angeles County Museum of Natural History in the United States.

Burd reviewed the data to 1992, and again to 1998, independently of prior analytical approaches (Burd and Ellis 1994, Burd 1999). She started by graphing scatter plots of likely interrelated data (e.g., the number of species versus distance from the discharge point), then used appropriate similarity and statistical testing to separate effects by near-, mid-, and far-field analyses.

This chapter summarizes environmental impact on the sediment benthos to 1998, the level and distribution of that impact, ecosystem recovery from the impact, and the natural variation in the biodiversity at sampling (reference) stations remote from the tailings. The main statistics used are numbers of different species and the identifications and numbers of the most abundant species. Ellis (2000) offers a more detailed species analysis, and the data set can also be obtained from ICM on disk. ICM hopes to make the data set available through the World Wide Web in some form. Annual reviews of environmental effects and recovery during and following mining are available in the mine's Annual Environmental Reports (ICM 1970–1999).

Direct viewing and photography from the submersible *Pisces IV* (Ellis and Heim 1985) and scuba dive surveys (see Chapter 9) have confirmed the presence of organisms on the tailings.

THE SEDIMENT HABITAT AND BIOLOGICAL INTERACTIONS

Although many physical and chemical parameters of the sediment habitat affect the infaunal benthos, typically only a few parameters are actually measured at any site during infaunal biodiversity surveys. Of these, grain size is most often documented, as a means of describing the habitat in categories of gravel, sand, and mud (silt and clay—the terminology of Folk 1974—is used in this chapter). There may also be measures of organic loading and trace metals where these are site-significant. At ICM, trace metals (Cu, Mo, and Zn) were routinely measured and reported annually as potential indicators of contamination and tailings deposition. Cu in particular has been routinely considered, along with tailings thickness measures, as an indicator of the rate and pattern of tailings deposition throughout the inlets (see also Chapters 5 and 10).

It should be noted that for the infaunal benthos, grain-size data beg the question of pore space characteristics, for which the conventional geophysical measure of porosity would be a supplemental indicator. The infauna inhabits the pore spaces and may change pore space size and pattern by their activities (their bioturbation). Also pore space characteristics are affected by the geophysical phenomenon of compaction. In short, pore space may be more important than grain size, but it is very rarely documented.

At ICM, grain size was initially documented by a Ph.D. thesis research study conducted from 1970 to 1973 (Johnson 1974). Grain size was also documented in 1972, 1997, and 1998, from the annual mine surveys carried out in March of each year for sediment trace metal analyses (ICM 1973 and 1999). Other grain-size analyses were performed in 1972 and 1973 (Jones 1974) and 1975 (Goyette and Nelson 1977). A second Ph.D. thesis research study (Hay 1981) furnished further details on grain size around the tailings density current in Rupert Inlet. Finally, there were two post-closure grain size data sets from the benthic sampling stations in 1997 and 1998 (ICM 1999). In all, several hundred samples have been analyzed for grain size throughout the fjord complex during a period of 29 years. Table 8.1 summarizes the data for samples at or near the benthic sampling stations.

The preoperational data from 1970 and 1971 (Johnson 1974) came from 41 samples at or near 21 of the routine sampling stations. Before mining began, surficial sediments almost everywhere in Rupert Inlet contained a sand component ranging from silty sands to sands (at and near station 13). In the fjord basin between Rupert and Holberg (station 9) Inlets, the strong tidal jets had removed fines to leave sands and gravels. In Holberg Inlet from midway between stations 9 and 6 to station 2, grain structure was finer, ranging down to silty clays, except at station 24 where sand reappeared. These are all partly fluvial sediments that have been deposited and reworked in the fjord system since it was opened up to the sea through Quatsino Narrows in the postglacial period. In the context of the infaunal benthos, Holberg Inlet sediments were typical deep fjord, fine-grained deposits, whereas the deeper parts of Rupert Inlet were unusually coarse-grained with a sand component, especially from station 13 to station 9 and slightly beyond. This is presumably an

TABLE 8.1 Sampling station sediment designations based on grain size (Sheppard or Wentworth terminology as used by original authors)

Sampling Stations	Pre-discharge 1970–1971	Early Years 1972–1975	Detailed Surveys 1977 and 1979	Postclosure 1997 and 1998
		Holberg Inlet		
24	Sand			
1–6	Silts with clays	Silts with clays	—	Silts with clays
		Coal Harbour		
7 and 8	Silts with sand	Sands with silts	—	Silts with sand
		Rupert/Holberg		
9	Gravels and sands	Silts to sands	Sands with some silt	Fine sand
		Varley Bay		
10 and 11	Sands, with gravel and silt	Sands with silt	—	Sands with silt
		Rupert Inlet		
12, 14–20	Silts, with sands and clays	Sands with silts and clays	Silts, rarely with sand	Sands to silts
13	Sands	Sands with silts and clays	Silts with some sand	Fine silt 1977 Fine sand 1998
21	Sand and silt	Sands with silt	—	Sand with silt
		Quatsino Sound		
22	—	Fine sand	—	Fine sand
23 and 25	—	Fine sand	—	Fine sand

effect of the periodic deep tidal jet (see Chapters 4 and 6). Benthic communities on both coarse- and fine-grained habitat have been documented and described from the general region of Vancouver Island (Ellis 1967, 1968a and b, 1971, 2001) and are available for comparison with those in the Rupert/Holberg fjord complex.

The tailings produced by the milling process at ICM had a median grain size of approximately 50 μm, which would produce a deposit of coarse silt by the Wentworth scale (Folk 1974). However it is to be expected that heavier, coarser grained particles would settle first nearest the tailings outfall, and that finer grained particles would disperse and settle farthest away from the outfall. The tailings at the discharge point were therefore coarser than sediments in Holberg Inlet, but finer than those deeper in Rupert Inlet and in the Rupert/Holberg basin.

The thickest tailings, beyond the capability of the 60-cm sampling device (a Phleger corer) to measure, quickly (1–2 years) extended from the outfall to the south shore of Rupert Inlet (station 17), down the fjord trough to the deep basin between Rupert and Holberg Inlets (station 9), and eventually (by 1988) onto station 6 in Holberg Inlet. The bulk of the tailings were within the deep targeted area below 100 m, but some drifted up into Holberg Inlet (e.g., station 4 and even station 2), apparently only intermittently as occasional clouds of particles. Some also upwelled into shallow water in the Hankin Point area, and settled in shallow water there, or flowed out into Quatsino Sound.

Sediment grain-size data are available from the early years of mine operations in March 1972, from December 1972 quarterly to September 1973, and from 1975 from several sources (ICM 1973–1974, Johnson 1974, Jones 1974, and Goyette and Nelson 1977). Thus data are available from 6 months after tailings discharge started and before the tailings had deposited at measurable levels beyond the Rupert/Holberg deep basin. Table 8.1 summarizes data that can be assigned to the benthos sampling stations. There is considerable variability between surveys, but summarizing maps are available (Johnson 1974, Figures 48 and 55). By 1973, the maps showed the coarsest deposits of >80% sands in a strip along the fjord trough at station 13 down to station 9, including a coarse tailings fraction of <6 ϕ. The Rupert/Holberg basin had dramatically changed from sands and gravels preoperationally to finer sands, including silt-sized tailings particles (7–8 ϕ). Goyette and Nelson (1977) confirmed by grain-size analyses that some of the upwelling tailings in the Rupert/Holberg basin had deposited in the shallows in and around Hankin Point. A coarse tailings fraction (<6 ϕ) was by the outfall, as expected, but the finest deposits (<20% sand) are upfjord from stations 15 and 16 with ϕ values of 6–7 (~0.010 µm, fine silts). These must be the finest fraction of the tailings (mean 50 µm, ~4.3 ϕ). Tailings had barely penetrated into Holberg Inlet (below the 100-m target depth), and as they were very fine-grained (like the natural sediments), they did not substantively change the grain size of the habitat. In terms of the infaunal habitat, by 1973 there had been substantial change in the Rupert/Holberg basin in particle size (i.e., reductions from gravel and sand to finer sands), but few grain-size changes were seen elsewhere. The maps in Johnson (1974) probably oversimplified the early changes, as the next paragraph shows.

Hay (1981) provides grain-size data from detailed surveys of the tailings beds in Rupert Inlet from December 1977, and from February and August 1979. The very high variability in grain size among the three surveys is a very unexpected finding. Table 8.1 summarizes the variability at or near the routine sampling stations. In detail, station 15 ranges from a fine sand to a coarse silt, station 9 from a coarse sand to a coarse silt. Even more revealing are Hay's Figures 46 and 50, which show the percentage of isolines for percentage of sand content. In February 1979 sandy patches were scattered down the line of tailings flow, but these had largely disappeared by August 1979. Evidently, the tailings beds do not present a uniformly fine-grained habitat for colonizing benthos but are continually being physically reworked (see Chapter 4). The dynamics affect stability (erosion and resettling) and grain size (silts to sands), thus affecting biodiversity. At any one place, at times the habitat approximated the predischarge condition with some sand, but at other times it was a much finer grained deposit. Instability continued as the habitat changed. This varying situation was apparently the equilibrium condition during mining operations.

The 1997 and 1998 data were collected in March of both years, 15 and 27 months, respectively, after the mine was closed in December 1995 (Table 8.1). The central Rupert/Holberg Basin (station 9) up Rupert Inlet to station 13 was then a fine sand, much finer than before mining, although it had shown even finer deposits at times during mining. From right after mine closure to date, Rupert Inlet has remained largely silts, thus becoming finer through tailings discharge. Holberg Inlet has remained as fine grained as it was both before and during mining. At a distance of 10+ km from the outfall there has been no substantive physical change in the habitat; all such physical changes were limited to the target area from Rupert Inlet to the Rupert/Holberg basin. Burd and Ellis (1994; Figure 22) plot tailings thickness against distance from the outfall and show a rapid reduction in deposits at a distance between 7 and 8 km.

Shallow water stations 1 (Holberg) and 21 (Rupert) did not have their habitat changed by the tailings, and in Quatsino Sound, neither did stations 22 and 23.

The question from these habitat descriptions is whether the grain-size changes, whatever their causes, have much impact on the biodiversity. Ellis (2000) has shown that there have been biodiversity losses under thick and rapid tailings deposition, but that a sustained succession in terms of number of species, numbers of organisms, and primary opportunists was established within 1–2 years of mine closure. The issues are whether the species composition has changed because of changed sediment type, whether it will change further, and whether this is significant to the ecosystem and its resource use.

First, the persistent crab fishery indicates that at least one upper level predator (Dungeness crab; see Chapter 10) that feeds on sediment benthos maintained itself (in spite of a commercial fishery). As it is an omnivorous feeder, it appears that benthic species compositional changes may not be important to it within the known levels of 50%–70% change from time to time and from place to place (Burd 1999). In any one year, approximately 300 species were documented, and through the 25 years of monitoring, about 1,000 species were present at one time or another.

Second, it is well established that gravel, sand, and mud habitat have markedly different and characteristic faunas (e.g., Thorson 1957). At station 13 the ICM data show that a characteristic sand fauna was appearing in 1997 and 1998 in association with the coarsening of the sediment. However, an article by Wu and Shin (1997) indicates that grain size alone may not be as definitive a determinant of eventual species composition as conventionally expected. Presumably other habitat factors related to, or independent of, grain size and compaction are also determinants; e.g., organic loading, permeability, and pore space characteristics, and the particular primary opportunists that settle first with their specific bioturbation patterns. The composition of a biological community at any time is a consequence of its history (Matthews, Landis, and Matthews 1996). Such dynamics of community succession can be modeled as "space-time worms" (Landis, Matthews, and Matthews 1996).

As of 2000 it appeared that the fine-grained sediments now widely spread in the Rupert/Holberg trough will gradually coarsen through whatever oceanographic dynamics created the originally coarser sediments there. However, the present erodibility of the sediments affected by the deep tidal turbulence will eventually lessen through compaction and at the fringes of the turbulent areas through the development of biofilms (Sutherland, Amos, and Grant 1998; Sutherland, Grant, and Amos 1998). Such biofilms have been routinely and widely observed and recorded in the staff's field notes during the benthic surveys as surface oozes. The reduced erodibility of the sediments will increase their stability, speeding the successional changes to the eventual equilibrium communities. There will inevitably be species-associated changes from the present state. The issue then becomes whether these changes and any temporary setbacks resulting from seabed instability will have ecosystem consequences. This issue is explored in the rest of this chapter.

THE DIVERSITY OF SPECIES (SPECIES RICHNESS)

At ICM, the number of different species, rather than conventionally calculated diversity indices, can be used as a species-richness measure because sampling procedures were identical from year to year. For this reason, variability in sampling procedures could not affect the numbers of species collected. This frees the data from errors in the use of diversity indices documented by Wu (1982), which arise from variable levels of identification (i.e., to species or to higher order taxa). The only major change in procedures with ICM's data was that the specimen sorters and their supervisors became more proficient with experience, resulting in some increases in the numbers of species counted over the years of the surveys. This means that the numbers of species before mining (1969–1970) and after mining (1996–1998) cannot be compared. In addition, the generally larger numbers of species in the final 10 years of sampling cannot be taken to mean a greater biodiversity than before mining.

The overall diversity of species is shown in Table 8.2 by the numbers of different species from 1970–1998 for all sampling stations. As explained previously, the values for 1973–1977 are omitted because of their inaccuracy. Table 8.2 also gives mean depths and Cu values, tailings deposition data, and some summary statistics—number of surveys at each station, the mean number of species for all years surveyed, and the mean number from 1993 to 1997 (or 1998).

Table 8.2 shows at which stations (e.g., stations 15–17) there were major decreases in the number of species (and when), which reveals the impact of tailings deposition. These stations coincide with the area of thickest and quickest tailings deposition (Figure 8.2). In this area, tailings deposits reached >60 cm (the length of the sampling corer used to take samples for Cu analysis) within 1–2 years of mining initiation. Echo-sounding monitoring showed that at each of these thick-tailing stations, after a further 5–10 years depending on location, water depth eventually decreased as a result of tailings deposition of up to approximately 50 m (e.g., station 9). Chapter 4 describes seismic surveys that report similar deposition and depth reduction levels.

Stations 15, 16, 17, 13, and 9 (ordered by distance from the discharge point) are within the thick tailings area. Table 8.2 shows that these stations, particularly 15 and 16, showed major reductions in biodiversity through losses in numbers of species. Decreases started within 3–9 months of the mine's opening. But the organisms decreased to zero (i.e., total obliteration) only twice at the time of annual sampling: at station 15 in 1982 and at station 16 in 1978. Either some species could survive rapid deposition by burrowing up, or recolonization was starting quickly, before the next annual survey.

Generally major decreases were to <10 species; i.e., some organisms could survive the rate of deposition or could colonize within a year. For each year in which there was a major decrease, the numbers recovered to some extent over the next few years. This appears to be derived from the dynamics of the tailings density current, which by changing its flow rate left areas of tailings to stabilize. Stable tailings beds allowed organisms to settle and grow. Only station 15, closest to the discharge point, had species numbers <10 during most of the mine's years of operations. Burd (1999) plots the number of species against the distance from the outfall and shows that most sampling stations with <10 species were within 9 km of the outfall.

Numbers of species were restored to levels (generally >30 species) within the range of those outside the thick tailings area by September 1996, 9 months after mining ceased, except for station 16, which required another year. However, numbers had started to increase in September 1994 before mining operations began to wind down. This appears to have occurred following a slight change in the outfall discharge point in 1994, so that the tailings density current, which kept the monitored sites unstable, was no longer near the sampling stations.

Initial data interpretation (1971–1975) in the then new field of monitoring seabed impact from an industrial waste was substantially incorrect. This was due in part to the low numbers of specimens reported as experience was gained with implementing the surveys and processing the samples, and due in part to misinterpretation. As late as 1974, the data from the most affected stations were still being interpreted as meaning obliteration

TABLE 8.2 Number of species at each sampling station each year per unit area sampled (0.15 m^2)

	Holberg Inlet										Rupert Inlet												Quatsino Sound				
Stn. No.	24	1	2	3	4	5	6	7	8	B	9	10	11	12	13	14	15	16	17	19	20	21	22	23	25	38	39
Depth (m)*	75	19	98	76	117	128	137	29	26	159	149	26	18	25	135	73	51	91	95	72	40	10	23	110	169	188	184
Cu (ppm)*	92	46	183	122	241	332	320	262	140		487	207	100	275	514	521	3,217	492	355	415	231	69	181	231	151	66	80
Tailings (cm)*	0	0	26.5	23	23.5	23	13	7.0	5.0			8.5	2.5	6.0		46				36.5	10	0	5.5	4	3	0	0
(year)			1991	1994	1989	1987		1987	1992			1984	1979	1987		1978				1992	1994		1990	1993	1993		
Surveys†																											
January 1970		19	11	12	12	8	9	16				20		10	16		16	11	15	21	18		21				
June 1971		41	7	16	12	20	26	29	35		43	29	25	31	4	28	28	13	32	19	26	30	43	43			
Tailings discharge started October 1971																											
December 1971		61	17	22	21	19	35	43	31		25	55	41	37	29	53	29	15	74	31	46	52	64	72			
June 1972		24	15	11	10	12	17	20	41		13	34	31	31	11	38	1		14	23	32	26	39	19			
December 1972		35	10	13	18	18	22	29	22		21	32	43	25	5	14	3		13	15	27	28	41	50			
1977	20	24	19	15	16	19	17	23	35		12	29	27	18	15	16	4	4	4	17	18	20	25	43			
1978	20	30	13	16	12	17	25	27	29	21	13	18	22	23	7	16	5	0	1	13	21	25	31	27			
1979	19	31	13	15	19	23	15	29	31	17	13	27	26	25	8	23	4	3	9	10	23	27	42	47	29		
1980	18	31	22	17	15	17	18	33	38	18	22	35	40	32	1	28	9	8	12	13	27	31	40	38	15		
1981	36	22	17	9	12	24	23	40	37		10	25	42	35	9	17	1	3	16	13	29	60	44	74	57		
1982	30	21	25	16	19	23	25	42	45	11	5	38	46	33	4	19	0	5	11	20	32	53	63	47	57		
1983	45	41	27	10	23	24	33	55	43	12	13	37	56	32	4	30	2	2	6	15	37	50	64	83	54		
1984	40	43	28	29	31	37	33	54	37	11	29	61	54	41	14	40	2	4	4	19	35	61	69	84	52		

(Table continues on next page.)

TABLE 8.2 Number of species at each sampling station each year per unit area sampled (0.15 m^2) (continued)

	Holberg Inlet										Rupert Inlet												Quatsino Sound				
Stn. No.	24	1	2	3	4	5	6	7	8	B	9	10	11	12	13	14	15	16	17	19	20	21	22	23	25	38	39
1985	43	42	24	23	28	25	27	41	54	10	12	46	40	39	8	27	11	4	6	11	28	49	55	70	49		
1986	41	76	19	26	25	39	36	72	63	20	32	90	92	45	20	42	8	2	4	14	32	102	93	107	63		
1987	48	78	25	31	28	42	20	88	77	8	22	99	84	66	11	49	3	9	13	25	50	103	83	121	59		
1988	64	85	28	34	25	32	18	89	75	13	26	76	78	48	18	49	6	10	15	30	54	83	121	142	77		
1989	53	89	26	19	25	23	27	61	52	15	18	48	93	39	5	35	7	10	15	19	52	87	116	139	72		
1990	58	123	34	24	31	28	28	85	73	27	32	69	77	64	14	35	5	8	15	15	49	99	119	152	99		
1991	54	88	27	31	28	39	27	69	73	8	8	64	95	59	8	32	3	15	14	22	68	77	100	132	126		
1992	55	79	31	42	30	26	25	81	59	10	20	71	78	71	10	37	8	10	9	6	46	76	107	123	121		
1993	54	81	32	24	31	30	22	62	57		16	55	76	51	14	36	4	8	9	6	42	67	118	123	115	160	67
1994	46	75	20	37	20	25	27	56	42		25	48	67	61	22	38	17	13	24	6	57	63	83	153	87	117	82
1995	41	76	35	34	39	40	38	64	48		38	63	97	82	24	48	34	15	41	20	69	105	145	114	98	107	106
Tailings discharge ceased December 1995																											
1996	62	97	34	39	41	29	23	97	79		71	127	137	70	57	61	42	15	42	26	65	102	139	161	96	201	132
1997	82	151	41	55	48	49	48	88	113		54	96	119	88	49	56	28	25	43	40	75	129	162	177	123	171	129
1998			52								81				70		59	58									
No. of surveys	21	26	27	26	26	26	26	26	25	14	25	24	24	24	25	24	25	23	24	24	24	24	26	25	19	5	5
Means—all	44	60	22	24	24	26	26	54	52	14	26	56	65	46	15	35	9	8	17	18	42	66	76	92	76	151	103
Means—1993+	57	96	32	38	36	35	32	73	68		41	78	99	70	33	48	25	15	32	20	62	93	129	146	104	151	103

* Depth and Cu values are means from annual surveys (Ellis 1999). Tailings are maximum increases over 1 year with the year identified (Ellis 1999).
† 1973–1976 species data were incomplete so are omitted.
Notes: From 1977, annual surveys were in September each year. Blank spaces mean no survey, and 0 means that zero species were collected at that sampling station during that survey.

of the fauna throughout the tailings area (ICM 1975, Goyette and Nelson 1977). We were not expecting, and thus missed, what in hindsight was obviously visible in the data as early as 1972. We did not then see that the low numbers of species present could mean that recolonization was occurring (ICM 1973) following obliteration. In 1975 this misunderstanding of continuing biodiversity obliteration on the tailings was corrected (Ellis 1975), but it still recurs in the literature from time to time (e.g., Childerhose and Trim [1979]). In 1999, the author verbally corrected a statement made about benthic obliteration at ICM at a public meeting on submarine tailings placement. The outdated information is likely to continue to appear and will need to be corrected each time.

In situ experimental colonization tests were reported by Taylor (1986) and summarized in Ellis (1989). Species richness in experimental trays increased within 1 year to levels shown by the field samples (although with species differences). Settlement on the seabed was greatest after the summer larval dispersal period.

Similarity analyses of the species and abundance data each year since 1986 when they were initiated (ICM 1987–1999), and subsequently by Burd and Ellis (1994) and Burd (1999), have shown that the deep remote sampling stations in Holberg Inlet and Quatsino Sound tended to be more similar to each other than to the affected stations in Rupert Inlet. The remotest deep stations are 39, 38, and 25 in Quatsino Sound and stations 24, 2, and 3 in Holberg Inlet. Table 8.2 shows that the number of species in all these stations can be 100 or more, but even in the past 10 years (of greatest sorting efficiency) numbers could nevertheless drop to as few as 20 (e.g., station 2 in 1994). Burd (1999) has shown that variability from year to year in these remote stations could vary from 50%–70% (see also the section on similarity analyses later in this chapter).

These remote deep stations (39, 38, 25, 24, 2, and 3) can be taken as the reference stations for comparison in each year, and over the past 10 years, for comparison with the seriously affected stations (9, 13, 15, 16, and 17). Note that the intended before and after comparisons from pre-mining surveys are not available because sample sorters became more efficient.

The remaining deep stations (4, 5, 6, 19, and 23) have shown some species decreases, indicating some biodiversity losses, but these were always followed by a recovery within 1–2 years. Station B at the edge of the thickest tailings deposits is the extreme example. By 1996 and 1997 these stations at the edge of the thick tailings deposits had recovered their numbers of species. Note that Cu values >~100 ppm indicate some deposition of tailings (pre-mining ambient levels, and levels at remotest stations 38 and 39, are generally less than 50 ppm). Stations 2, 3, 25, and possibly 24 show some Cu elevation (to ~200 ppm). Evidently there is a level of copper increase (which in insoluble chalcopyrite is an expression of tailings deposition, not of Cu leaching potential from the tailings [Pedersen 1985; see Chapters 5 and 10]) that the infauna tolerates. This appears to be at least 200 ppm. Burd (1999) also indicates that particulate copper has not adversely affected the organisms. This level of 200 ppm should not be taken as a threshold to toxicity of copper in mine tailings, but as a surrogate measure of tailings deposition.

This in turn means that there is a rate of tailings deposition that the infauna tolerates, either by organisms burrowing up faster than the tailings deposits or by larvae settling rapidly as some organisms are smothered, leaving space in the tailings deposits for colonization.

The monitoring design included sampling stations in shallow water to check whether tailings were present in water depths less than the targeted deposition area of 100+ m. These are stations 1, 7, 8, 10, 11, 12, 20, and 21. Numbers of species are within the range of the deep reference stations. In general this is to be expected because the shallow water infauna tends to be more diverse than the deep water infauna. Cu values at these stations show some elevations, which indicates that some tailings were suspended into shallow water (stations 7 and 8 in Holberg Inlet; stations 11, 12, and 20 in Rupert Inlet; and station 22 in Quatsino Sound). There is no detectable impact on the infauna at these Cu levels, which are generally <200 ppm. It appears, then, that these shallow water areas also can tolerate some low rate of tailings deposition.

The raised Cu levels in shallow water at stations 7, 8, and 11 in the Holberg/Rupert complex, in station 22 in Quatsino Sound, and in deeper water in Quatsino Sound (station 23) reflect the resuspension and upwelling of tailings by tidal jets at the junction of Rupert and Holberg Inlets (Chapter 4) and redeposition in shallow water with some escape into Quatsino Sound (less than 0.4%; ICM 1980). There has been no detectable impact of these dispersed tailings at these sampling stations in water depths of <100 m and deeper in Quatsino Sound.

The rate of tailings deposition that the benthos can tolerate is an important management statistic, and some information is available from the species diversity numbers. Table 8.2 shows that at station 14 between 1977 and 1978, 46 cm of tailings deposited, and although numbers were already low (16 species), they did not decrease. Also, at station 19 between 1991 and 1992, 36.5 cm of tailings deposited with a drop of numbers of species from 22 to 6. At all other stations (e.g., 3, 4, and 5), in contrast, maximum annual changes of approximately 20 cm did not produce substantial decreases in species numbers. It appears that in this area infaunal benthos can tolerate deposits at the rate of 20–30 cm in 1 year.

This is not unexpected from data elsewhere with tolerance values (because of upward burrowing) between 4–30 cm of rapidly deposited sediment (Newell, Seiderer, and Hitchcock 1998). It is also a conservative estimate of tolerance. The erratic distribution of sudden increases of maximum deposition through the years suggests that deposition is patchy and quick in the more remote parts of the inlets. This must come from clouds of tailings rather than from deposition continuing at a constant average rate through a year from a widely spread and layered plume.

This derived tolerated deposition rate at ICM of ~20 cm/yr is much higher than the previous estimate of 1 cm/yr first hypothesized in 1992 (ICM 1993). The 1992 statistic was derived from the inability to distinguish—by similarity analysis—stations 8, 20, 12, 10, and 7 (subjected to 5, 2.5, 5, 11, and 16 cm, respectively, of tailings since discharge started 22 years before in 1971) from zero tailings stations 11, 21, 1, 24, 23, and 25. A tolerance of 1 cm/yr was thus provided as a very conservative value. It can now be seen as substantially lower (i.e., too conservative) than the values up to ~20 cm deposited in 1 year without impact on the biodiversity, and the even higher values from other sources.

THE NUMBERS OF EACH SPECIES (SPECIES EVENNESS)

The total number of organisms collected (of all species lumped together) is not a particularly informative statistic after the numbers of different species have been considered (as previously discussed). In the early stages of ecological succession, following habitat changes, a few opportunist species can be present in enormous numbers, giving values for total numbers of organisms close to, or even greater than, those of a more diverse community in equilibrium with its habitat (Pearson and Rosenburg 1978). The equilibrium concept (a varying biodiversity in a varying habitat), as opposed to the more commonly used term "climax community," implicitly expresses the high biodiversity variability that can be expected (as at ICM) from time to time and from place to place in a varying habitat.

More significant in the context of the seabed adjacent to ICM are the actual species present and their numbers per unit area. With more than 1,000 species collected from more than 2,000 samples (~26 × 3 samples for 27 years), and averaging 300 species per year, it is not possible to show all the species' numerical data here.

It has previously been shown (Ellis and Hoover 1990, Ellis 1998 and 2000) that the first species to colonize tailings (within 1–2 years) and to sustain their populations (primary opportunists) are polychaete worm species such as *Chaetozone acuta, Cossura pygodactylata, Euchone incolor, Tharyx multifilis, Lumbrineris luti,* and *Nephtys cornuta.* It was noticeable that the well-known primary opportunist *Capitella capitata* (Pearson and Rosenburg 1978) only occasionally colonized the tailings deposits (e.g., station 13 in 1990 at 2,000/m^2), and then did not sustain itself. This is presumably a result of the tailings' lack of the organic materials that are usually associated with *Capitella*'s colonizing; e.g., sewage, pulp mill, and other organically loaded deposits.

Four of the six polychaetes are particle-feeding sedentariate worms and are a mixture of suspension and deposit feeders (Conlan 1977, Roe 2000). Two, *Lumbrineris luti* and *Nephtys cornuta,* are mobile omnivores or carnivores. Secondary opportunist species appearing a year or so later are particle-feeding bivalve mollusks, especially the small clam *Axinopsida serricata* (Bright 1991), which can also assimilate dissolved organic matter. Numerical analyses had earlier indicated that many of these were ecologically significant in the region (Ellis 1969).

Ellis (2000) documented the following changes in species evenness:

- At affected stations 15 and 16 in Rupert Inlet from 1993 (the last year of severe impact) to 1998 (almost 3 years after mining ceased), the species listed previously regularly appeared from 1994—1 year after major defaunations at both stations—and then remained abundant. The secondary opportunist, *Axinopsida serricata,* appeared in 1994 and fluctuated in numbers thereafter. Another polychaete, *Ophelina acuminata,* also appeared in 1994 and then sustained itself but usually only in low numbers. *O. acuminata* had previously appeared in large numbers during the early 1970s but did not sustain itself (Jones 1974). It was then identified as *Ammotrypane aulogaster.*

- At two reference stations, 2 and 24, the primary and secondary opportunists frequently appeared in the top five species or slipped slightly below the top five as other species entered the fauna. It appears that the primary opportunists in this locality are not quickly displaced—if they are displaced at all—by later colonists, as can occur in some ecosystems.

- In spite of the lesser number of species at affected stations 15 and 16 compared with reference stations 2 and 24 (in 1997 28 and 25, respectively, compared with 41 and 82, respectively) there was a persisting similarity in the most abundant species among all four stations just before and after mine closure. The differences between the recovering and unaffected stations therefore lie in the greater diversity of relatively rare species. By 1997, fewer than 2 years after mine closure, the predominant species on recovering and unaffected stations were similar and sustaining themselves in both areas.

- The most abundant species in the two affected stations (9 and 13) closest to the deep basin at the junction of Rupert and Holberg Inlets have changed since 1995. This is the area where tidal jets have caused some resuspension and upwelling of tailings and caused, since mine closure, a gradual habitat change—the depth is increasing (from 120 m in 1995 to 140 m in 1998; see Chapter 4). The deposits are both coarsening (to fine sands) and becoming harder to penetrate by corer (compacting). The species in 1998 included the burrowing sea anemone *Halcampa decemtentaculata*, the sand-grain-tube-building polychaete worms *Owenia fusiformis* and *Pectinaria californiensis*, and several bivalve mollusks (*Macoma* spp. and *Yoldia scissurata*).

- The species evenness effects at the two stations where the habitat is changing physically from mud bottom to sand bottom following mine closure (lower Rupert Inlet to the junction with Holberg Inlet) are as follows: The area suffered species reductions during mine operations, but within 1 year after mine closure the increased biodiversity was showing changes to a sand bottom fauna. These changes are probably still in progress, but their endpoint in terms of the species making up a sand-bottom equilibrium community, the numbers present, and their variations from year to year and place to place cannot be predicted from the data available. However, it is probable that the total number of species and numbers of organisms will remain at the levels reached from 1996 to 1998.

- In spite of the dynamic habitat changes (such as grain sizes and instability), a consistent set of colonizing species was seen on both sands and silts following major species reductions (see previous). It appears that grain size was not a major determinant of colonization, but that a regionally determined array of colonizers settled on any stable deposit, at least within the range of sand to silt. This is important information for other locations where tailings may be deposited. Some local colonizers at ICM tolerated a substantial change in grain size between tailings and natural sediments, and the biodiversity is now changing with the habitat as it changes following mine closure. It seems possible that in accordance with the parallel level bottom community hypothesis of Thorson (1957), similar species (even from the same genera at ICM) will colonize mine tailings in many parts of the world.

- The species richness changes have had no effect on the stocks, and commercial fishery yields, of a high-level benthic predator, the Dungeness crab (*Cancer magister*).

THE BIODIVERSITY EXPRESSED IN TERMS OF HIGH-LEVEL TAXONOMIC UNITS

Certain types of organisms normally present in moderate to large numbers on mud and sand bottom were either absent from the recovering areas or present as only a few species (Ellis 2000). These include particularly echinoderms, such as brittle stars, sea cucumbers, and sea urchins, frequently present on fjord mud bottoms in western Canada (Ellis, 1967, 1968a, 1968b, 1969, and 1971) and easily identified. Evidently echinoderms are naturally few in species and numbers in deep water in the Rupert/Holberg complex, so they cannot be used as a bioindicator of recovery to the equilibrium community. In contrast, a higher diversity of crustacea was found in the reference areas than in the affected areas.

In summary, it appears that the mud bottom affected stations will sustain their succession to a stage where there is a greater diversity of crustacea than at present. This appears to be the only major biodiversity recovery change still in progress on the deep fjord mud seabeds 3 years after mining operations ceased. The time required for numbers of small crustacean species to increase to within the equilibrium range cannot be predicted from the data available.

SIMILARITY AND DIVERSITY INDICES

Similarity analyses were routinely implemented from 1986 using the Bray-Curtis coefficient and the SIGTREE software (Nemec and Brinkhurst 1988) for probability analysis (ICM 1987–1999). These first surveys used microcomputers with limited memory capacity, so a reduction procedure to ~100 species was developed by eliminating rare and ubiquitous taxa lacking distributional information (ICM 1986). In 1992 B. Burd was retained to review the data using a mainframe computer to process the data on all species (Burd and Ellis 1994). Her techniques were then used to analyze all data from 1994 on (ICM 1995–1999).

In all years there were distinct clusters of similar stations (severely affected stations 15–17, for example), but these were usually at low levels of similarity (e.g., <50%). It appears that similarities among and within stations changed greatly from year to year. This is a well-known phenomenon, first shown by Petersen and Jensen (1911). The phenomenon has been expressed quantitatively for the Rupert Inlet area (Burd 1999) as a normal similarity level of only 50%–70% (i.e., at any one station from year to year). Stations on similar habitats, such as shallow stations with few or no tailings (e.g., stations 1, 8, 11, 21, and 22) had equally low levels of similarity in any 1 year.

Diversity indices were also routinely calculated each year. These were the Keefe-Bergeson, Shannon, Pielou, and Margalef indices. These were also uninformative compared with the actual counts of species and numbers of organisms. Values of the indices for each station each year are available in the computerized data bank for further analysis if needed. The data bank also allows calculation of yet other diversity indices if these should ever be considered appropriate. In summary, the conventional biodiversity indices were calculated following current practice, but during the long term of 29 years, with a high natural variability of the infauna from year to year, they were relatively uninformative.

LARGE INFAUNAL SPECIES

The small sampling grab used in these surveys, which grabs a sample of area 0.05 m^2, and digs only to a depth of 5–10 cm, cannot collect large, deeply burrowed, or fast organisms. Fast organisms such as seabed-inhabiting fish tend to be sufficiently mobile that their growth is a function of a broad area. For this reason, they cannot be considered as characteristic of the area sampled by grab. In contrast, large sedentary organisms, whether at the surface of the sediments or burrowed in, are a product of the sampling site. These include organisms such as large clams (e.g., the commercial geoduck *Panope generosa*) and holothuria of several genera.

Records of large species seen on and off tailings during submersible dives have been documented by Ellis and Heim (1985); see also Chapter 9. About 20 species of fish and invertebrates, both mobile and sedentary forms, were seen on the tailings.

At ICM, a means was found to assess the entry of large, deeply burrowing but sedentary species into the biodiversity through the presence of juveniles in the (small) grab samples. The major forms are mollusks, so in recent years all mollusks were measured (length) and weighed. Generally, mollusks to ~3 cm in length were collected.

Juveniles of several of the larger organisms were collected, but not frequently (Ellis 2000). The two large local bivalve mollusks, *Tresus* and *Panope*, were not collected at all. They may be present in the fjord system, but if so they have not spawned with a widespread settlement, neither on nor off tailings during the monitored period. Occasional spawning (i.e., not annually), is not unexpected (Thorson 1957). The cockle *Clinocardium* appeared fairly regularly, including at station 7 with some tailings deposition (Table 8.2). The large mud bottom clam *Compsomyax* has appeared at station 17, where there had been severe faunal losses prior to 1993. Station 13, also with severe losses prior to 1993, has received settlements of *Clinocardium* and holothuria.

It appears that the system of identifying juveniles, at least to genera, can monitor the settlement of organisms that, when adult, are too large or too deeply burrowing to be collected by a small sampling device.

SUSTAINABILITY OF THE BENTHIC RECOVERY FROM TAILINGS IMPACT

The documented benthic recovery has been used to develop a set of criteria for measuring whether an ecological succession has started and is being sustained (Ellis 1998 and 2000). Sustained ecological succession implies that the succession will continue until a benthic community in equilibrium with the physical/chemical conditions has been attained. In practice, for the infauna, measuring whether a community is in equilibrium with its habitat is an impossible task because species diversity and evenness are so variable. This was well documented almost a century ago at the start of quantitative benthic biodiversity assessment (Petersen and Jensen 1911). However, it is relatively easy to document (as shown previously) that a consistent set of primary opportunists has settled, that the set is gradually being extended with later settling species, and that the range in numbers of species and their evenness falls within the range of the local equilibrium community. This state of sustainable ecological succession was achieved at all affected sampling stations except one (station 16) within a year of tailing discharge ceasing (by 1996). Station 16 had reached the sustainable succession stage a year later (Ellis 2000).

As described in the previous subsections, the marine benthos can now be expected to sustain itself indefinitely with additions of more species, especially crustacea. However, there may be temporary setbacks caused by erosion from tidal currents or slumping of perched deposits.

In conclusion, the data from ICM give a measure of sediment habitat changes from sands to silts and corresponding biodiversity levels. The data show immediate and slow losses in biodiversity (localized virtual obliteration on occasions) and quick (1–3 years) recovery to sustainable population levels on tailings. There was a consistent set of colonizing species, and similar species may colonize tailings elsewhere. Species compositional changes are still in progress on both sands and silts, but appear to have little significance to the ecosystem in view of the many species (more than 1,000 collected in 29 years of monitoring and ~300 in any 1 year) potentially able to populate the present range of coarse and fine-grained sediments in the inlet system, and the persistence of stocks of a commercially fished predator (Dungeness crab) in shallow and deep water throughout the inlets.

ACKNOWLEDGMENTS

I am very grateful for cooperation from many scientists and students over the years and for their ideas in benthic sampling and analyses. Particularly I thank B. Burd and students A. Jones, K. Conlan, L. Taylor, C. Heim, D. Bright, and G. Roe. Also, the data bank has been accumulated and documented by a series of staff members from ICM's environmental division and supervised over the years by C. Pelletier, R. Hillis, and I. Horne. Biologica Inc. performed the sample sorting, cleaning, and species identifications and counts since 1986 under the supervision of V. Macdonald, who has provided very valuable continuity in the species identifications since 1972.

REFERENCES

Anon. 1970. Public inquiry into the application from the Utah Construction and Mining Company. *Report of the Inquiry,* 2 volumes.

Bright, D. 1991. *Tissue Variability in the Infaunal Bivalve* Axinopsida serricata (Lucinacea:Thyasiridae) *Exposed to a Marine Mine-Tailings Discharge and Associated Population Effects.* Ph.D. diss. University of Victoria, Canada.

Burd, B.J. 1999. *Post-mining Recovery of Infaunal Benthos in Rupert Inlet.* Annual Environmental Report Vol. III, Part 2. Port Hardy, BC: Island Copper Mine.

Burd, B.J., and D.V. Ellis. 1994. *ICM Closure Plan: Review of Benthic Surveys 1970 to 1992 for Rupert/Holberg/Quatsino Inlet System.* Report to Island Copper Mine.

Childerhose, R.J., and M. Trim. 1979. *Pacific Salmon and Steelhead Trout.* Vancouver, BC: Douglas and McIntyre Press.

Conlan, K.E. 1977. *The Effects of Wood Deposition from a Coastal Log Handling Operation on the Benthos of a Shallow Sand Bed in Saanich Inlet, British Columbia.* M.Sc. thesis. University of Victoria, Canada.

Ellis, D.V. 1967. *Quantitative Benthic Investigations. II. Satelllite Channel Species Data, February 1965–May 1967.* Technical Report No. 35. Nanaimo, BC: Fisheries Research Board of Canada.

——. 1968a. *Quantitative Benthic Investigations. III. Locality and Environmental Data for Selected Stations (Mainly from Satellite Channel, Straits of Georgia and Adjacent Inlets), February 1965–December 1967.* Technical Report No. 59. Nanaimo, BC: Fisheries Research Board of Canada.

——. 1968b. *Quantitative Benthic Investigations. V. Species Data from Selected Stations (Straits of Georgia and Adjacent Inlets), May 1965–May 1966.* Technical Report No. 73. Nanaimo, BC: Fisheries Research Board of Canada.

——. 1969. Ecologically significant species in coastal marine sediments of Southern British Columbia. *Syesis* 2:171–182.

——. 1971. A review of marine infaunal community studies in the Strait of Georgia and adjacent inlets. *Syesis* 4:3–9.

——. 1975. Pollution controls on mine discharges to the sea. *Proceedings International Conference on Heavy Metals in the Environment, Toronto, Canada, October 27–31, 1975.* 677–686.

——. 1989. *Environments at Risk.* Heidelberg, Germany: Springer-Verlag. 329 pp.

——. 1998. Ecological criteria for determining the sustainability of restoration. *Proceedings Helping the Land Heal Conference, Victoria.* 133–137.

——. 1999. Station-Time Series Summaries 1970–1998. *1998 Annual Environmental Assessment Report.* Vol. III Benthos. App. 4. Port Hardy, BC: ICM.

———. 2000. Effect of mine tailings on the biodiversity of the seabed: Example of the Island Copper Mine, Canada. In *Seas at the Millennium*. C. Sheppard, ed. Vol. 3, Chapter 23. The Netherlands: Elsevier. 235–246.

———. 2001. *A Review of Marine Biological Data Assembled for the Proposed Gas Pipeline Through Satellite Channel*. Report to B.C. Parks. 15 pp.

Ellis, D.V., and C. Heim. 1985. Submersible surveys of benthos near a turbidity cloud. *Marine Pollution Bulletin* 16(5):197–204.

Ellis, D.V., and P.M. Hoover. 1990. Benthos recolonizing tailings beds in British Columbian fjords. *Marine Mining* 9:441–457.

Ellis, D.V., and V.I. Macdonald. 1998. Rapid preliminary assessment of seabed biodiversity for the marine and coastal mining industries. *Marine Georesources and Geotechnology* 14(4):1–13.

Ellis, D.V., T.F. Pedersen, G.W. Poling, C. Pelletier, and I. Horne. 1995. Review of 23 years of STD: Island Copper Mine, Canada. *Marine Georesources and Geotechnology* 13(1&2):59–100.

Ellis, D.V., and G.W. Poling, eds. 1995. Submarine tailings placement. *Marine Georesources and Geotechnology* 13(1&2). 233 pp.

Ellis, D.V., and J.D.R. Robertson. 1999. Underwater placement of mine tailings: case examples and principles. In *Environmental Impact of Mining Activities*. J. Azcue, ed. Chapter 9. Heidelberg, Germany: Springer-Verlag.

Ellis, D.V., R. Samoszynski, and A.A. Jones. 1991. Re-analysis of species associational data using bootstrap significance tests. *Water, Air and Soil Pollution* 59:347–358.

Folk, R.L. 1974. *Petrology of Sedimentary Rocks*. Austin, TX: Hemphill Publishing. 182 pp.

Gaston, K.J., and J.I. Spicer. 1998. *Biodiversity. An Introduction*. Oxford, UK: Blackwell Science. 113 pp.

Goyette, D., and H. Nelson. 1977. *Marine Environmental Assessment of Mine Waste Disposal into Rupert Inlet, British Columbia*. Fisheries and Environment Canada Surveillance Report EPS PR-77-11. Vancouver, BC: Environmental Protection Service. 91 pp.

Hay, A.E. 1981. *Submarine Channel Formation and Acoustic Remote Sensing of Suspended Sediments and Turbidity Currents in Rupert Inlet, B.C.* Ph.D. diss. UBC, Canada.

Holme, N.A., and A.D. McIntyre, eds. 1984. 2nd ed. *Methods for the Study of Marine Benthos*. Oxford, UK: Blackwell Science. 387 pp.

ICM. 1970–1999. *Quarterly and Annual Environmental Reports*, 1–4 volumes annually. Port Hardy, BC: ICM.

———. 1980. *A Study of the Suspended Sediment Movement Through Quatsino Narrows*. Port Hardy, BC: ICM Environmental Department.

Johnson, R.D. 1974. *Dispersal of Recent Sediments and Mine Tailings in a Shallow-Silled Fjord, Rupert Inlet, British Columbia*. Ph.D. diss. UBC, Canada.

Jones, A.A. 1974. *Effects of Mine Tailing on Benthic Infaunal Composition in a B.C. Inlet, with Special Reference to Sampling, Instrumentation and the Biology of* Ammotrypane aulogaster (Polychaetea; Opheliidae). M.Sc. thesis. University of Victoria, Canada.

Jones, A.A., and D.V. Ellis. 1976. Sub-obliterative effects of mine-tailing on marine infaunal benthos. *Water, Air and Soil Pollution* 5:299–307.

Landis, W.G., R.A. Matthews, and G.B. Matthews. 1996. The layered and historical nature of ecological systems and the risk assessment of pesticides. *Environmental Toxicology and Chemistry* 15(4):432–440.

Matthews, R.A., W.G. Landis, and G.B. Matthews. 1996. The community conditioning hypothesis and its application to environmental toxicology. *Environmental Toxicology and Chemistry* 15(4):597–603.

Nemec, A.F.L., and R.O. Brinkhurst. 1988. Using the bootstrap to assess statistical significance in the cluster analysis of species abundance data. *Canadian Journal Fisheries and Aquatic Science* 45:965–970.

Newell, R.C., L.J. Seiderer, and D.R. Hitchcock. 1998. The impact of dredging works in coastal waters: a review of the sensitivity to disturbance and subsequent recovery of biological resources on the sea bed. *Oceanography and Marine Biology: An Annual Review* 36:128–178.

Pearson, T.H., and R. Rosenburg. 1978. Macrobenthic succession in relation to organic chemical enrichment and pollution of the marine environment. *Oceanography and Marine Biology. Annual Review* 16:229–311.

Pedersen, T.F. 1985. Early diagenesis of copper and molybdenum in mine tailings and natural sediments in Rupert and Holberg Inlets, British Columbia. *Canadian Journal Earth Science* 21:1–9.

Petersen, C.J.G., and P.B. Jensen. 1911. *Valuation of the Sea. I. Animal life of the sea bottom, its food and quantity*. Report Danish Biological Station No. 20. Copenhagen, Denmark: Danish Biological Station.

Roe, G. 2000. *A Survey and Analysis of the Colonization and Succession of Marine Benthic Species on Mine-Tailings Deposits from Island Copper Mine, British Columbia*. Course ER390 Report. University of Victoria, Canada.

Sutherland, T.F., C.L. Amos, and J. Grant. 1998. The effects of buoyant biofilms on the erodibility of sublittoral sediments of a temperate microtidal estuary. *Limnology and Oceanography* 43(2):225–235.

Sutherland, T.F., J. Grant, and C.L. Amos. 1998. The effect of carbohydrate production by the diatom *Nitschia curvilineata* on the erodibility of sediment. *Limnology and Oceanography* 43(1):65–72.

Taylor, L.A. 1986. *Marine Macrobenthic Colonization of Mine Tailings in Rupert Inlet, British Columbia*. M.Sc. thesis. University of Victoria, Canada.

Thorson, G. 1957. Bottom communities (sublittoral or shallow shelf). *Memoirs Geological Society of America* 67:461–534.

Wu, R.S.S. 1982. Effects of taxonomic uncertainty in species diversity indices. *Marine Environmental Research* 6:215–225.

Wu, R.S.S., and P.K.S. Shin. 1997. Sediment characteristics and colonization of soft-bottom benthos: A field manipulation experiment. *Marine Biology* 128:475–487.

Underwater Biodiversity Surveys and Biological Colonization of the Waste Dump Shoreline

Derek V. Ellis

INTRODUCTION

The environmental and biological impact from tailings in shallow water, and recovery from that impact, has been observed and documented by a series of dives (both scuba and submersible) starting before mining began in 1970 and ending 3 years after mine closure in 1999. The dives showed that limited amounts of tailings upwelling into shallow water had little impact on the algal forests of the underwater cliffs and reefs and on their dependent epifauna. The tailings in shallow water also settled in roughness features of the underwater cliffs, creating sediment patches a few centimeters deep. These deposition beds appeared to be highly dynamic, coming and going with the strong tidal currents of the area. The biodiversity in these algal forests and sediment patches remained high throughout the area during mining operations, although species changes caused by tailings deposition and its subsequent dispersal were observed. The deepest sections of the underwater cliffs, largely limited to the junction of Rupert and Holberg Inlets, were buried by up to ~50-m depth of tailings deposited there so that the rock face epifauna at those depths >~100 m was changed to a sediment infauna. In 1999, $3\frac{1}{2}$ years after mine closure, the biodiversity was still high, and biodiversity changes following tailings dispersal away from the shallow area appear to be ongoing. The shoreline of the waste-rock dump has been monitored since 1996 to check the pattern and rate of biological colonization. The expected biodiversity of the worldwide rocky shore zonation patterns appeared and became sustained within 2 years.

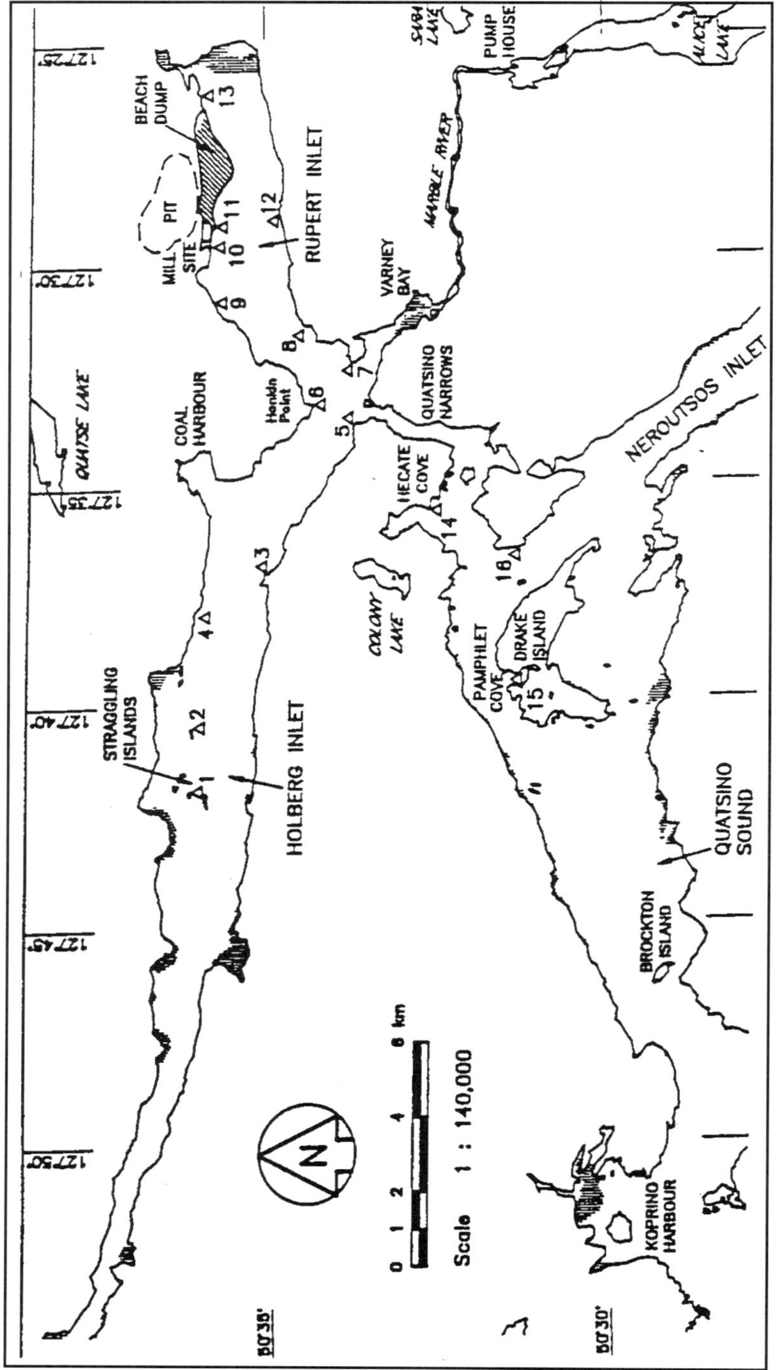

FIGURE 9.1 Scuba dive transects were located at the settlement plate sites (see Chapter 10), including the mill dock (#10) and the tailings outfall (#11)

THE UNDERWATER SURVEYS

As explained in Chapter 8, monitoring of the biodiversity of shallow water algal forests, of sediment patches on the underwater cliff sides, and of changes induced by settlement of tailings was not undertaken initially because tailings were expected to remain below the discharge depth of ~50 m. This situation changed when it became apparent in April 1972 (Ellis 1972) that repeat transmissometry surveys of water turbidity (preoperational survey September 21–22, 1971; postoperational surveys October 8–23, 1971; November 30–December 2, 1971; and April 9–15, 1972) showed that tailings were upwelling periodically in the Hankin Point area.

By 1973, the upwelling surges of tailings were sufficiently visible at the surface in the Hankin Point area and drifting seaward through Quatsino Narrows to merit detailed observations of tailings upwelling and transport (see Chapter 4; Goyette and Nelson 1977). From 1974 to 1984, ICM, government agency, and university scientists engaged in diver or submersible observations or both and reviewed the data obtained. There was a major thrust to develop quantitative procedures to better establish the level of impact in shallow water.

Essentially, biodiversity remained high in the algal forests during mine operations whether tailings were present or not. In 1999 scuba divers obtained detailed photographic and videotape documentation that showed the high levels of biodiversity on the three sites surveyed that received tailings: Hankin Point, the mine dock, and along the tailings outfall. The details are given in the sections that follow.

Original data from the surveys and limited distribution reports about them are archived at the University of Victoria library. Report summaries are available through <http://gateway3.uvic.ca/archives/featured_collections/mesc/home.html> or by sending an e-mail to dvellis@uvic.ca.

INITIAL DIVER SURVEYS BY MINE ENVIRONMENTAL STAFF AND CONSULTANTS, 1972–1974

Scuba diver surveys by ICM environmental staff and consultants to check tailings presence and their impact in shallow water began in 1974 (Dobrocky Seatech Ltd. 1974, Foreman 1974 and 1976). The surveys were generally conducted at the sampling stations for measures of suspended solids (Chapter 2) and biological settling on colonization plates (Figure 9.1 and see Chapter 10).

The last of the three surveys was conducted under rigorous position and depth-finding conditions by R. Foreman (UBC Botany Department), a specialist in the then-developing techniques of diving surveys for biodiversity assessment. The three surveys yielded reports on the biodiversity to ~20-m depth on hard rock underwater cliff surfaces, on sediments close by or in patches on shelves along the underwater cliffs, and on pilings at the concentrate loading dock at the mine.

The first two surveys in 1974 established that there was a rich biodiversity in the algal forests growing on the cliffs and a rich biodiversity also on the adjacent sediment beds, including areas with visibly obvious tailings.

Foreman's (1976) multivariate statistical analysis of the final 1974 survey data, based on samples identified under laboratory conditions, allowed him to draw conclusions such as (direct quotes are provided in places):

- The differences among 10 sites were largely natural. Only two of the sites, within ~2.5 km of the mine dock and tailings outfall, possibly showed effects from mine operations, but those effects were also possibly induced by the mine's waterfront activities rather than tailings.

- The number of species and number of organisms were high at the two sites by the mine (e.g., 104 and 87 species of algal macrophytes, respectively). These numbers were actually higher than at some of the reference sites, but there appeared to be some impoverishment impact with depth.

- "The nearshore benthic communities will undergo modifications but will not be eliminated by mine activities except locally around the waste dump."

- "The changes should stabilize, with constant mine input, in two to five years from present except for the continual substrate modification by sediment deposition."

- "If the community is modified to the point where overstory species are excluded this will be accompanied by a considerable change in species composition, with greater development of the understory species."

- "If sediment deposition at any site becomes deep enough to mask the minor (3–10 cm) substrate roughness features there will be a pronounced change in biotic recruitment patterns, species composition, and biomass when compared to present levels."

- "Extensive sediment [tailings; author's clarification] would select for macrophyte genera adapted to soft substrates and an increase in *Neogardiella*, *Gracilaria*, and *Zostera* would be expected. Sedimentation rates do not appear sufficient to bury these species and their performance would be expected to improve as they extend upwards into areas from which they are presently excluded by substrate."

- "Changes in the macrophyte community composition will be accompanied by changes in the faunal community."

Foreman examined the data and drew conclusions that any substrate change (i.e., tailings replacing natural sediments or depositing in patches on surface roughness) would be followed by adaptive species changes (i.e., replacement of rock face algal communities by sediment algal communities). His conclusions could be checked visually by the routine servicing (four times annually) of the settlement trays (see Chapter 10). As a result, the mine made the decision not to routinely monitor the algal forests by diver but to repeat diver-based surveys at intervals. The first of these repeats took place in 1983 as described in the following section.

DIVER SURVEYS BY FEDERAL GOVERNMENT SCIENTISTS, 1970–1975

Federal government scientists (Goyette and Nelson 1977, Petrie and Holman 1983) conducted seven sets of diver surveys before mining began and during the early years of mining operations (1970–1975).

In August 1972, Varney Bay near the mine site (Figure 9.1) was showing no signs of tailing infiltration, nor was Holberg Inlet in October 1973.

By April 1974, a series of dives showed fresh sediments, probably tailings, at and near Hankin Point. By the mine site itself there were substantial fresh deposits, probably of mud from waterfront activities. From May 1974 to July 1975 there were increases in sediment loadings at Hankin Point and nearby, in the form of widespread films of deposits on algae, and in patches on cliff roughness features. Thickness of the patches was recorded as 2–20 cm. These films and patches could be seen extending onto shore at extreme low tides.

In 1978 a federal-provincial government inquiry to review the significance of the environmental changes to that time (see Chapter 2 and Waldichuk and Buchanan 1980) encompassed the diver and shoreline observations. Tailing patches exposed at extreme low tides were supporting beds of eelgrass and other sediment-inhabiting species (Buchanan 1982). The maximum thickness reported was 15 cm. The patches were accepted as highly dynamic, depositing, resuspending, and redepositing in the highly variable tidal currents in the area, but maintaining a biodiversity as Foreman (1976) predicted.

SUBMERSIBLE SURVEYS, 1975–1984

Three series of submersible surveys were conducted during the mine's operations, all using the federal government's submersible *Pisces IV*. Government scientists (Goyette and Nelson 1977, Petrie and Holman 1983) made six descents in the fjord complex between February 11 and 14, 1975. *Pisces IV* was made available in 1983 and 1984 to the University of Victoria marine research program for further dives in the Rupert Inlet area—on April 13, 1983 (eight dives—Ellis 1983), and May 1–2, 1984 (four dives—Ellis and Heim 1985)—for development of underwater quantitative survey techniques.

The 1975 government dives showed that when the submersible was on the seabed at 90 and 75 m depth to the east of the mine site, turbidity was so great that nothing was visible. Turbidity increased rapidly below 15 m depth. In Holberg Inlet, at 115 and 80 m, depth tailings (gray) were visible on the seabed, but there were light brown sediments (probably natural) in shallower water at 40–65 m depth.

The 1983 and 1984 university dives concentrated on developing a methodology for quantitative observations (Ellis and Heim 1985). After a preliminary dive in an area to determine the depth at which visibility was reduced to zero, the submersible was moved into shallower water nearer shore to dive at just above that depth. Observations of the seabed were made at 13–15 m, 16–18 m, and 27–53 m depths. Some tailings (gray sediments) were present at those depths in Rupert Inlet, supporting stands of burrowing anemones, starfish, crab, and shrimp. Burrow holes of unidentified organisms were also seen. Two observers recorded quantitative data, counting within a 1-m^2 quadrat suspended in front of the submersible, and along a transect measured by an anchored, spooling-out, 100-m-long rope marked at 5-m intervals (Plate 9.1). The three dives in 1984 were between shoreline and three of the routine benthic monitoring stations (stations 20, 17, and 14) where the biodiversity was measured each year (see Chapter 8).

The submersible dives from 1975–1984 agree in showing that within the tailings plume on the seabed, biodiversity measures cannot be obtained visually. In shallow water to 50-m depth with fewer tailings, a visible, substantial biodiversity of large, benthic organisms could be seen living on and burrowed into the sediments.

DIVER SURVEYS, 1983 AND 1999

In 1983 and 1999, mine environmental contractors conducted extensive diver surveys to assess and photograph rock face and sediment biodiversity. The surveys were conducted by experienced underwater photographers and biologists.

The 1983 surveys were undertaken on March 17, June 20, and September 16 (DeLisle 1983). The divers descended vertically from the boat to the anchored sediment traps and colonization plates at 20-m depth, then ascended by transect to shore, photographing en route. A rich biodiversity of large algae and epifauna was seen at all sites, including Hankin Point and the mine dock.

In 1999 (July 12–17), more than $3\frac{1}{2}$ years after mine closure in December 1995, a final extensive diver survey was undertaken (Lacasse 1999). Dive transects were made at seven sites, including Hankin Point, the mine dock, and down the tailings outfall to ~20-m depth (Figure 9.1). As in all previous surveys, there was a rich biodiversity of large algae and epifauna, including on sediment deposits, which at the deeper parts of the transects in Rupert Inlet had received some tailings. Plates 9.2–9.7 illustrate the biodiversity in Rupert Inlet, on tailings, and in unaffected areas, $3\frac{1}{2}$ years after mine closure. Approximately 200 photographs and a videotape were obtained. The organisms present are similar to those documented elsewhere throughout the fjord system.

CONCLUSIONS DRAWN FROM THE UNDERWATER SURVEYS

Diver and submersible observations (some with samples and counts) and photographs show that from 1972 (1 year after mine operations started) to 1999 ($3\frac{1}{2}$ years after mine closure), a substantial biodiversity was maintained in the algal forests and on sediment deposits with and without tailings throughout the Rupert Inlet fjord complex.

The first detailed impact assessment in 1974 by Foreman (1976) predicted future changes that were confirmed by DeLisle (1983) and Lacasse (1999). Essentially, Foreman predicted that the biodiversity of rock cliffs (where subjected to tailings deposition) would change to those species adapted to inhabiting sediments. Goyette and Nelson's dive reports (1977) are not inconsistent with Foreman's predictions.

Currently (in 2001), if the tailings deposits in shallow water at Hankin Point are being eroded and dispersed in this high-energy area (see Chapter 4), it is to be expected that the algal forests and their dependent epifauna will revert to being more like those at the reference areas with little sediment. Where tailings remain overlain by natural sediment deposits—for example, off the tailings outfall site—the algae, epifauna, and infauna can be expected to remain similar to the present; i.e., showing a high biodiversity of sediment-adapted species.

There has been no quantitative estimate of the proportion of the deepest underwater cliff face buried by tailings deposits in the deepest parts of the inlet system, where Rupert and Holberg Inlets meet. The bottom depth there was raised by ~50 m (see Chapter 4), which means that the deepest cliff faces and their epifauna were buried (but not the algal forests, which cannot exist at those depths). Lesser depth increases elsewhere mean lesser similar burial. The contour map of the original inlet system (Chapter 4) indicates the relatively small amount of such deep underwater cliffs. It also appears that, since mine closure in 1995, the water depth to tailings in the deepest basin has increased (as a result of erosion and dispersion). This suggests that at least some of the buried underwater cliffs will be re-exposed, leading to recolonization by epifauna.

SHORELINE COLONIZATION OF THE BEACH WASTE DUMP

The Shoreline and the Beach Dump

Because the expected tailings impact was below the discharge depth of 50 m, the intertidal zone was not selected in 1970 for regular monitoring. After tailings were recognized as upwelling into shallow water at Hankin Point (Chapter 4), this decision was reconsidered. It was then rejected again, in favor of periodic shallow water scuba diver surveys. Initial dive surveys (see previous section) and viewing from boats had shown that there was abundant biodiversity within the kelp forests on the underwater cliffs (in spite of the tailings present on fronds and suspended in the water), and also on and in the tailings that had deposited in shallow water at Hankin Point. It was believed, then, that there would be even less impact on shoreline biodiversity, and that monitoring was unnecessary.

The intertidal biodiversity again became an issue as the beach rock-waste dump (Chapter 1) was built out into Rupert Inlet (Figure 9.1). Casual observations showed that where the dump face stabilized, intertidal growth of rockweeds, mussels, and starfish, among others, quickly started the characteristic worldwide pattern of vertically layered biological zones between low and high tide (Stephenson and Stephenson 1949 and 1972).

Finally, the regulatory agencies required ICM to shape the face of the beach dump into a series of shallow water embayments that would function as nursery areas for Pacific salmon fry and smolts on their outward migration from spawning streams to the open coast. At this time, the decision not to monitor the intertidal biodiversity was reversed. It was decided to monitor the intertidal biodiversity each year after mine closure to determine if, how, and when the biodiversity was established on the beach dump face. This, along with results from shallow water (10–20-m depth) infaunal assessment (Chapter 8), was to be a surrogate measure of the ability of the shallow-water embayments to function in providing feedstock for young salmon. The biodiversity of the embayments was not monitored directly.

Several factors played into this decision. First, the intertidal biodiversity monitoring could be undertaken expeditiously with reporting each year within a few days of completing the field surveys. This could be accomplished by drawing on the Stephensons' global zonation concept (see following section). Second, the known enormous variability of infaunal (sediment-inhabiting) benthos meant that any shallow-water monitoring (1–10 m) would need to establish the natural range of variability over, for example, a 5-year period, using the time-consuming, precise, species-identification procedures followed for the sediment infauna throughout the inlets (Chapter 8). This would substantially delay expeditious reporting because a rapid monitoring protocol had not then been developed. In addition, the extensive fjord infaunal monitoring (Chapter 8) had shown that from 10–20-m depth (even where there had been disturbance from the deposition of some tailings), the biodiversity remained substantial. The expectation has been that if both 10–20-m infaunal monitoring and intertidal biodiversity monitoring show normal levels of biodiversity, the immediate (1–10 m) shallow-water biodiversity will also be within the normal range.

The Assessment Working Concept: The Stephensons' Global Rocky Shore Zonation Scheme

The Stephensons (1949 and 1972) have documented globally that rocky shore biodiversity falls into a vertically layered pattern of zones consisting essentially of (Plate 9.8):

- **The Uppermost Supra-littoral Fringe**–includes a black belt of a maritime lichen *Verrucaria* with a zone, apparently of bare rock, below. The zone actually supports abundant small periwinkles (genus *Littorina*) and cropping growths of microscopic (invisible) green algae. Some limpets and other organisms may be present, especially where sheltered from exposure by cracks in the rock surfaces.

- **The Middle Littoral Zone**–a zone of abundant biodiversity, which is specifically adapted to the alternating sea and air exposure in the intertidal. Usually there is an uppermost brown band set by rockweeds of the genus *Fucus*, with underlying dense beds of barnacles of several species. Nearer low tide the biodiversity varies greatly from place to place and time to time, but dominant species and keystone species are often mussels and starfish, respectively. (Dominant species are those that create habitat, whereas keystone species are herbivores and carnivores that exert population control on their feedstocks.) The habitat of this zone is therefore biologically created in three dimensions by the growths of algae, barnacles, and mussels. Many small species (especially salmon-feedstock crustacea) live on and in this habitat, retreating between the dominant growths at low tide, especially during daytime low tides.

- **The Lowermost Infra-littoral Fringe**–the zone where occasional very low tides expose shallow-water species that can tolerate some exposure to air. Kelps of many species dominate the zone, and there is a diverse epifauna and other algal growths (red, green, and brown algae, along with marine grass beds).

Some of the Stephensons' original work was undertaken on Vancouver Island (Stephenson and Stephenson 1961), and the regional intertidal has been well-documented by, for example, Carefoot (1977); Kozloff (1983); O'Clair and O'Clair (1998); and Ricketts, Calvin, and Hedgpeth (1968). Application of the concept to the ICM area is therefore based on considerable knowledge of the regional intertidal zonation, some of it near the mine site (Cross and Ellis 1981).

The monitoring concept was designed to determine if, how, and when the Stephensons' zones were established on the beach dump face, allowing for the dump face to consist of boulders and stones, not solid rock face. The expectation was that algal growth and dependent epifauna would be patchily distributed along the zones because of their reliance on the presence of large, relatively immovable boulders. The surveys were undertaken by walking transects down the beach dump face, annually during summer low tides (June–August in any 1 year), and documenting the relative abundance of organisms selected as being the local zonation-indicator forms. The results were documented each year by tables and photographs (see Ellis 1998 and 1999).

Shoreline Zonation on the Beach Waste Dump (Plate 9.8)

Engineering of the beach dump face ended in April 1996, and the first intertidal survey was conducted in September 1996. Thereafter, annual surveys were conducted in June, July, and August at the time of very low tides.

In 1996 it was shown that Stephenson-type zonation was present along the beach dump face except in the easternmost section, where there were indications of a zonation start but not of fully developed zonation. In this eastern section, engineering had ceased only 4 months before. A distinct biological front separated the two sections (Plate 9.8, bottom).

By 1997, almost all the beach dump face was showing the Stephenson zones, and it was determined that where the zonation had been established by 1996, it had been sustained into 1997. The biological front between the two differing sections in 1996 had almost disappeared by 1997. The farthest eastern section of the face, including one embayment, was under impact from a log-handling operation and was not responding to the same extent. After a similar result in 1998, this area was removed from the mine's monitoring.

Further surveys in 1998 and 1999 confirmed the sustainability of the zonation pattern and documented the manner in which the species varied from year to year. In 1998 there had been a dense settlement of barnacle larvae overgrowing and preventing other rock-attaching forms (rockweeds, mussels, etc.) from settling. In 1999, the barnacles had decreased to normal levels, allowing other forms to colonize, some of which were patchily abundant in some areas and not present in others (e.g., isopods, cockles, and fish [eelpouts]).

This Stephenson-type procedure for intertidal surveys has merit in that it eliminates the need to establish the natural levels of variability of the equilibrium community in terms of species present (species diversity) and their abundances (species richness). Instead, the procedure establishes that the globally characterizing bands of organisms are visibly present, with local (but unidentified) indicator species.

In conclusion, the sustained intertidal biodiversity in the normal Stephenson pattern on the beach dump face, combined with the shallow water (10–20-m depth) faunas in Rupert Inlet being within the normal range of biodiversity, is taken to mean that the immediate shallow-water biodiversity off the beach dump face is normal (in terms of providing feedstock for young salmon on outward migration).

ACKNOWLEDGMENTS

I am very grateful to many marine scientists and mine environmental staff members over the years for their cooperation when I was involved in some of the fieldwork for this chapter. I much appreciate the work of the scuba divers and either their analyses (R. Foreman) or the photographs and videos that they took and reported (R. DeLisle and S. Lacasse). I also thank my professional associate and wife K. Ellis for her participation in the intertidal biodiversity surveys from 1996–1999.

REFERENCES

Buchanan, R.J. 1982. A government inquiry into mine waste disposal at Island Copper Mine. In *Marine Tailings Disposal.* D.V. Ellis, ed. Chapter 7. Ann Arbor, MI: Ann Arbor Science.

Carefoot, T. 1977. *Pacific Seashores: A Guide to Intertidal Ecology.* Seattle: University of Washington Press. 208 pp.

Cross, S.F., and D.V. Ellis. 1981. Environmental recovery in a marine ecosystem impacted by a sulfite process pulp mill. *Journal Water Pollution Control Federation* 53(8):1339–1346.

DeLisle, R. 1983. *Underwater Photographic Survey #1.* Port Hardy, BC: ICM.

Dobrocky Seatech Ltd. 1974. *Transmissometer Field Notes and Station Records. Diving Survey Results.* Port Hardy, BC: ICM. 32 pp.

Ellis, D.V. 1972. *Interpretation and Evaluation of Transmissometer Surveys Undertaken by Dobrocky Seatech Ltd. on Behalf of Utah Construction and Mining Co. Ltd. in the Area Adjacent to Island Copper Mine.* Report to the University Independent Agency, ICM Environmental Control Programme. Vancouver, BC: UBC. 61 pp.

——. 1983. *Submersible Surveys in the Tailings Receiving Area for Island Copper Mine, April 13, 1983.* Victoria, BC: University of Victoria.

——. 1998. *The Beach Dump Face at Island Copper Mine: Intertidal Biodiversity Colonization and Sustainability 1996–1998.* Port Hardy, BC: ICM.

——. 1999. *Reclamation of the Beach Dump: The 1999 Biodiversity Survey of the Shoreline Face.* Port Hardy, BC: ICM.

Ellis, D.V., and C. Heim. 1985. Submersible surveys of benthos near a turbidity cloud. *Marine Pollution Bulletin* 16(5):197–204.

Foreman, R.E. 1974. *June Reconnaissance of Benthic Sublittoral Communities in Rupert and Holberg Inlets.* Report to the University Independent Agency. ICM Environmental Control Programme. Vancouver, BC: UBC.

——. 1976. *Nearshore benthic biota study in Rupert and Holberg Inlets.* Report to the University Independent Agency. ICM Environmental Control Programme. Vancouver, BC: UBC. 63 pp.

Goyette, D., and H. Nelson. 1977. *Marine Environmental Assessment of Mine Waste Disposal into Rupert Inlet, British Columbia.* Surveillance Report EPS PR-77-11. Vancouver, BC: Environmental Protection Service. 93 pp.

Kozloff, E.N. 1983. *Seashore Life of the Northern Pacific Coast.* Seattle: University of Washington Press. 370 pp.

Lacasse, S. 1999. *1999 Rupert Inlet Underwater Photographs.* Port Hardy, BC: ICM.

O'Clair, R.M., and C.E. O'Clair. 1998. *Southeast Alaska's Rocky Shores: Animals.* Duke Bay, AK: Plant Press.

Petrie, L., and N. Holman. 1983. *Pisces IV Submersible Dives 1973–1982.* Regional Programme Report 83-20. Vancouver, BC: Environmental Protection Service. 314 pp.

Ricketts, E.F., J. Calvin, and J.W. Hedgpeth. 1968. 4th ed. *Between Pacific Tides.* Stanford, CA: Stanford University Press.

Stephenson, T.A., and A. Stephenson. 1949. The universal features of zonation on rocky shores. *Journal Ecology* 37:289–305.

——. 1961. Life between tidemarks in North America. IVa. Vancouver Island I. IVb. Vancouver Island II. *Journal Ecology* 49:1–29; 227–243.

——. 1972. *Life Between Tidemarks on Rocky Shores.* New York: W.H. Freeman. 425 pp.

Waldichuk, M., and R.J. Buchanan. 1980. *Significance of environmental changes due to mine waste disposal into Rupert Inlet.* Vancouver, BC: Fisheries and Oceans, Canada, British Columbia Ministry of Environment. 56 pp.

Fisheries, Tailings Bioassays, Trace Metal Bioaccumulation in Benthos, and Settling Plates

Derek V. Ellis

INTRODUCTION

Crab populations maintained themselves and supported a commercial fishery throughout mining operations, including over the tailings beds. Although salmon populations were not measurable, the mine supported a salmon hatchery that has added substantially to local populations of juvenile salmon being released to the sea. Tailings bioassays from the undiluted discharged effluent almost always gave a 100% 96-hour survival of test fish. Trace metal bioaccumulation appears to have occurred only on mussels that inhabited the ship-loading dock by the mine. The trace metal bioaccumulation appears to have resulted from fugitive dust during loading operations, not from bioactivation of metals from discharged tailings. In particular, annual monitoring has shown no biomagnification of trace metals up the food chain by the species considered most likely to show such biomagnification (the local stock of the scavenger/predator Dungeness crab). Settling plates measuring settlement of tailings and colonization of marine larvae showed the presence of tailings in shallow water in Rupert Inlet, but without reductions in larval colonization.

This chapter reviews monitoring data accumulated on fishery species (crab and salmon), on the tailings bioassays, on the bioaccumulation of trace elements, and on the biological colonization of in situ settlement plates.

Fishery species are considered first. The two types (crab and salmon) reported were selected for specific reasons. First, government-produced fishery statistics for all locally fished species refer to catch areas far broader than those affected by the mine's operations. For this reason, the government statistics could not reflect the impact, if any, of the mine on the crab and salmon stocks. This rendered the statistics unusable as monitoring data.

Of the various fishery species, the crab stock could be monitored by the mine's environmental staff by setting traps identical to those used by the fishing fleet, under the guidance of a local commercial fisherman. The crab data, therefore, reflect the crab fishery catches during mine operations and since mine closure.

The local salmon (Pacific salmon of the genus *Oncorhynchus*) could not be monitored by methods simulating those of the commercial fishing fleet, so fishery yields could not be monitored. This applied even when a commercial fishery (for pink salmon *O. gorbuscha*) was allowed in Rupert Inlet near the mine site—the fishing boats did not have to report their catches in such a small area, but their data were factored into the broader catch-area statistics. In addition, an annual test fishery by the mine, even with catch replacement, would probably have endangered the local stocks. As a result, surrogate measures were developed to determine the state of the salmon stock or data were extracted from government annual spawning stream surveys. Because the latter gave a strong indication of continuing decline in the local salmon stocks starting well before mine operations, the mine took the initiative, with other local interest groups, to develop a salmon hatchery. The yield and return statistics from this hatchery offer the best continuing surrogate measure that could be obtained of the state of the salmon stocks in the area. The hatchery continues its operations now after mine closure.

Original data from the surveys and limited distribution reports about them have been archived at the University of Victoria library. Report summaries are available through <http://gateway3.uvic.ca/archives/featured_collections/mesc/home.html> or by sending an e-mail to dvellis@uvic.ca.

THE CRAB FISHERY AND YIELD

Because it can be collected in a standardized fashion by a trap fishery as previously explained, Dungeness crab (*Cancer magister*) was selected for monitoring. The species was also useful as a monitor of the risk of trace element biomagnification up the seabed food chain in Rupert/Holberg fjords, as individuals are resident predators feeding on the locally available benthos. If these feedstock species should bioaccumulate trace toxins, the predatory Dungeness crab in the fjord would risk suffering from the effects of biomagnification.

The monitoring was conducted throughout the three inlets, Rupert, Holberg, and Quatsino, either three or four times each year—in March, June, September, and December. Figure 10.1 shows the sampling stations. Two standard commercial traps were set for 18 hours (overnight) at each station. They were baited with fish offal similarly to the commercial fishery. All specimens were identified to species (*C. magister* or *C. productus*), measured (carapace width), weighed, and sexed. Three legal size Dungeness crabs were retained for trace element analysis; the rest were released. The commercial traps used are designed to allow small crabs less than 14-cm carapace width to escape. As a result, size data are biased against small crabs and are not considered here.

Table 10.1 reports the most important results. In all years, crabs were caught in the tailings areas in Rupert/Holberg Inlet. The commercial fishery continued throughout the inlets during mine operations and has done so since mine closure. Monitoring was stopped in 1998, $2\frac{1}{2}$ years after mine closure.

Rupert Inlet has catches intermediate in value between Holberg Inlet and Quatsino Sound. The catch data fluctuate from year to year, but not in a way that bears a relationship

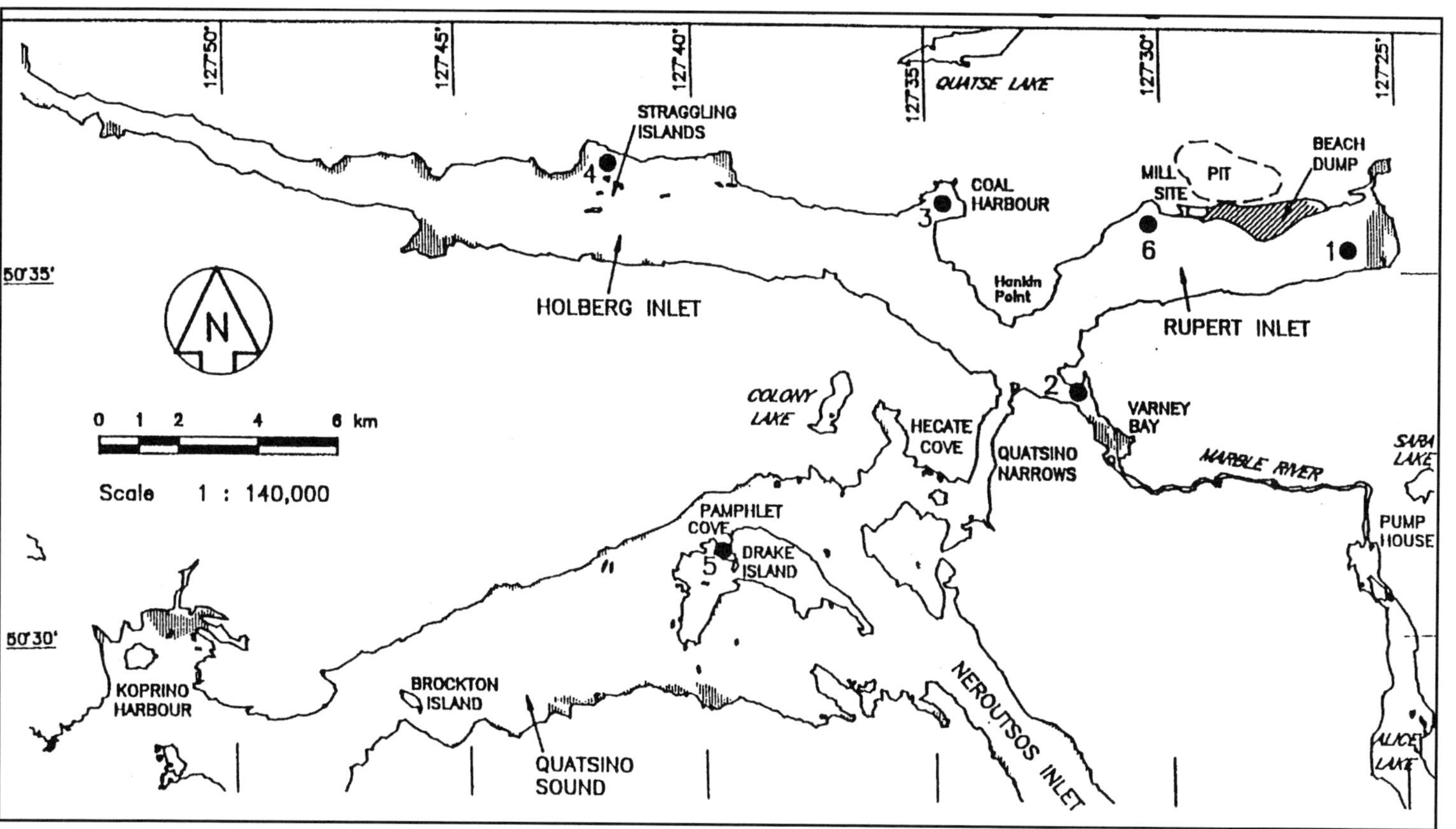

FIGURE 10.1   Map showing crab sampling stations

TABLE 10.1   Summary of catches of Dungeness crab (*Cancer magister*), 1970–1998

Year	Rupert	Holberg	Quatsino
1970	22.8	19.5	24.5
1971	18.5	21.2	18.9
1972	11.5	14.9	13.8
1973	13.1	23.8	6.8
1974	19.4	22.8	9.8
1975	34.6	41.0	8.3
1976	28.2	30.8	16.8
1977	25.4	34.7	2.9
1978	24.6	31.4	15.3
1979	35.1	38.4	14.4
1980	25.5	27.3	4.9
1981	21.3	26.0	6.3
1982	19.4	22.4	4.6
1983	20.2	26.1	7.3
1984	14.5	20.9	6.4
1985	16.1	19.8	4.5
1986	11.2	21.2	4.9
1987	17.3	23.9	9.8
1988	15.4	26.3	6.5
1989	5.0	18.7	5.0
1990	16.9	18.5	4.1
1991	24.8	35.1	4.9
1992	27.7	29.3	4.8
1993	26.6	25.4	12.0
1994	21.3	25.0	6.8
1995	18.5	25.3	10.9
1996	15.6	19.3	3.8
1997	7.7	5.5	3.0
1998	10.2	5.7	6.6
Mean	19.6 (7.3)	24.1 (8.0)	8.6 (5.3)

Notes: Values are mean numbers/trap for an 18-hour overnight set. Each area has 2–3 catch sites. Mean values with standard deviations. 1970–1971 are preoperational catches, 1972–1995 are during operations, and 1996–1998 are post-closure catches.

with mine operations (although at the time, the decline in Rupert Inlet from 1971–1973 caused considerable alarm).

The lowest value in Rupert Inlet in 1989 has no obvious cause and was accompanied by low values throughout the inlets that year. Similarly, the low 1997 and 1998 values are area-wide and not mine-related because mine closure was in 1995.

Trace metal monitoring has been documented in the annual environmental reports (ICM 1970–1999) and is reviewed in the following text. At no time was there indication of biomagnification of trace metals through the Dungeness crab route. As the most likely route for biomagnification, because both crabs and feedstocks are resident in the area, the absence of biomagnification was taken as an indicator that trace element biomagnification did not materialize as a risk during mine operations.

In summary, the crab fishery was maintained during mine operations. The mine's monitoring data show that crabs were available throughout all inlets with fluctuations from year to year unrelated to mine operations, and that crabs, even the stocks closest to the mine site in Rupert Inlet, were not contaminated by trace elements during mine operations.

SALMON AND THE SALMON HATCHERY

The stocks of Pacific salmon (genus *Oncorhynchus*) have been of environmental concern to ICM since its original environmental plans were developed. The submarine tailings disposal system was intended to avoid affecting the salmon-spawning streams near the proposed mine site by placing the tailings deep in the sea below the shallow-water migratory routes of young salmon out to sea. Placing the tailings on land would have had an impact on salmon- and trout-spawning streams adjacent to the mine.

To recap, ICM did not conduct its own monitoring program on salmon returns for various reasons. First, the federal Department of Fisheries routinely monitors the numbers of salmon breeding in spawning streams. Second, conducting a test fishery even with catch replacement would probably have endangered the local small stocks. The Department of Fisheries' own catch statistics were from too large an area to reflect runs returning to the Rupert/Holberg/Quatsino fjord complex. This applied even when the occasional fishery (for pink salmon *O. gorbuscha*) was permitted in the fjords. It should be noted that the Department of Fisheries required no remedial action for the salmon stocks during mine operations.

However, ICM took two voluntary steps for salmon conservation.

A review of Department of Fisheries information about the salmon stocks was undertaken in 1974 (Ellis and Jewsbury 1974). At that time it was concluded that as many as 100,000 salmon or more could migrate through Rupert Inlet to the spawning streams in any year, and that as many as 7.5 million fry migrated outward in the spring of each year. Government spawning stream surveys indicated that there had been a continuing decline in the local stocks throughout the previous decades. Following the review, ICM undertook some beach seining and echo sounding of the outward juvenile migration each spring. The intent was to confirm that migration took place in shallow water above tailings discharge depth, and this was verified by each year's monitoring.

The second step was to initiate and support a salmon hatchery on a spawning river draining into the fjords. ICM supported this effort both financially (half the operating costs) and in kind (e.g., provision and maintenance of electricity supply, pumps, and land by the mine's water-extraction pump house). The site selected was on the Marble River (Figure 10.1) just below Alice Lake. This is a fair-sized river with stocks of all salmon species and unknown numbers of trout.

The hatchery is supported by ICM, Western Forest Products, and the North Vancouver Island Salmon Enhancement Association. During the past 18 years, returns of chum (*O. keta*), chinook (*O. tshawytscha*), and coho (*O. kisutch*) salmon have averaged 1,000–3,000 annually, and returns of sockeye (*O. nerka*) and pink (*O. gorbuscha*) salmon have averaged 100–500 annually. Releases of fry from the hatchery into the river have been into the hundreds of thousands. For example, 647,996 chinook fry were released in 1993. The number varies greatly from year to year and species to species. A

1998 supporting letter from the Friends of the Marble River notes "Since 1980 the hatchery has released over 37,000 steelhead [trout, *O. mykiss;* author's clarification], 700,000 coho and 5.2 million chinook fry. Chinook runs in the Marble have risen from a low of 500 fish prior to the enhancement effort to an average of 3,000 and [sic] still rising."

Salmon were not required to be analyzed for trace elements because, for example, the period during which these migrants were exposed to mine wastes in the sea was too short to expect any bioaccumulation. Salmon fry and smolts from Rupert Inlet pass through the inlet in a matter of days each spring, and in shallow water above the discharge depth of 50 m. After a period of 2 to more than 4 years at sea, returning adults may spend some time (days or weeks) at depth in the inlet, but because they are not feeding, they cannot absorb trace elements through the food chain. There was no indication (see Chapter 6) that water column trace elements were raised in ways in which the fish could directly absorb them (through, for example, their gills). In short, the regulatory authorities considered that salmon were not at risk from trace element contamination.

In summary, salmonids were not affected by the submarine tailings disposal system adopted as a substitute for on-land placement, which would have affected some of the nearby salmon- and trout-spawning streams. ICM took the initiative to develop and support a salmon hatchery, which has been effective in reducing the long-term continuing decline in local salmon-spawning stocks.

TAILINGS BIOASSAYS

The mine's various discharge permits required it to undertake acute toxicity tests of its tailings to fish, and to demonstrate that it was meeting the regulatory agency requirement of 100% survival (with an allowance for occasional 50% survival because test fish can die for reasons other than the test materials). BC Research Corporation, an independent agency with considerable experience in bioassays, conducted the bioassays. The testing procedures were originally from APHA, AWWA, and WPCF (1971) and were later set by Fisheries and Environment Canada (1977). The tailings sample for the bioassays was taken from the tailings effluent stream before discharge into Rupert Inlet.

Table 10.2 shows that bioassays using conventional LC_{50} 96-hour exposure tests of juvenile rainbow trout (*O. mykiss*), or juvenile coho salmon (*O. kisutch*), almost always gave 100% survival of the test fish during the 25 years of operations of the mine. At such times, then, no acutely toxic components were present in the effluent. Note that test results before July 1973 were not reported in the mine's quarterly monitoring reports (ICM 1970–1973), and appear to have been lost (or at least are not readily available).

On very few occasions from 1973 to 1995, there was survival less than the allowed 50%. These values occurred between 1976 and 1983 and have not recurred since. For 8 months from November 1978 to July 1979, there were erratic but dramatic reductions in survival down to 0%.

When the testing laboratory reported reduced survivals, ICM's procedure was to take further effluent samples and repeat the tests, as well as to check possible causes. In 1979, the problem appears to have been a new processing frother, which the mine introduced on the basis of satisfactory stand-alone bioassays by the manufacturer. The irregular toxicity was presumably caused by interactions of the frother with other processing chemicals. The new processing chemical was replaced by the original frother, and bioassay survival returned to 100%.

TABLE 10.2 Tailings bioassay results; acute toxicity tests (LC$_{50}$ 96-hour) on juvenile salmonids

Year	Number of Bioassays	Number of Tests with 100% Survival	Comments
1971			Data not found
1972			Data not found
1973	8	8	
1974	7	7	
1975	11	11	
1976	13	8	January–March 60%–0%, repeats 100%–90%
1977	14	14	
1978	13	12	One 0%, repeat 100%
1979	28	17	Eleven 80%–0% to July, repeats some 100%
1980	12	12	
1981	16	15	One 50%
1982	15	13	One 0%, one 50%, repeats 100%
1983	12	11	One 90%
1984	12	12	
1985	12	12	
1986	12	12	
1987	12	12	
1988	12	12	
1989	12	12	
1990	12	12	
1991	12	12	
1992	12	12	
1993	12	12	
1995	12	12	

This successful monitoring alert illustrates that the LC$_{50}$ 96-hour acute bioassay procedure worked very effectively as a monitoring system of the toxicity of the tailings effluent. In other words, interactions among individually nontoxic processing chemicals could be detected.

TISSUE MONITORING OF TRACE METAL BIOACCUMULATION BY BENTHOS

In 1969 and throughout the 1970s, there was considerable concern about the potential of the tailings to release biologically active toxic trace elements known to be in the ore, such as copper and arsenic. However, research in sediment geochemistry (Sly 1996) and monitoring of the sediment metal levels and interstitial studies at ICM (Pedersen 1984 and 1985, Pedersen and Losher 1988; see Chapter 5) gradually established that marine sediments in general and specifically at ICM were neutralizing and precipitating, rather than bioactivating, environments. In hindsight this is not surprising. The ICM copper source is low-grade (~0.5% copper) chalcopyrite (see Chapter 2), which is virtually insoluble. Also, seawater is neutralizing to acid sources, and fine-grained sediments are anoxic a few centimeters below the interface with the seawater (Pedersen 1985).

Consequently, although initially of concern, elevated copper values in the Rupert Inlet sediments, up from ~50 ppm to ~1,500 ppm (0.0015%) cannot be taken to indicate biologically available copper. Instead, they indicate areas where tailings have settled. Detailed analyses of the relationships between copper levels and seabed biodiversity showed that at sediment Cu values >700 ppm, species biodiversity declined (Burd and Ellis 1994, Burd 1999; see Chapter 8). This effect occurred in Rupert Inlet within ~5 km of the tailings outfall during mine operations. The dynamics of biodiversity losses and recoveries indicated that Cu values >700 ppm marked areas with dynamic settling, erosion, and resettling of tailings.

However, as a result of the concern about the bioactivation of toxic trace metals, their bioaccumulation by benthos has been monitored since 1971, just before discharge began. Various marine algae, marine grasses, marine benthic invertebrates, and fish species were monitored. Trace metals monitored were cadmium, copper, molybdenum, and zinc, plus the element arsenic. Trace metals in plankton were also monitored and consistently gave no evidence that these drifting organisms of the fjord ecosystem were bioaccumulating trace elements.

A logistical problem developed very quickly after monitoring started in that some of the initially selected species varied greatly in availability from year to year—for example, from zero present to juveniles only, to large specimens only, to mixed juveniles and large specimens, but not many of each. Dungeness crab was the potential biomagnifying species most consistently available in adult form (see previous text and Table 10.1 for data on its consistency). Some shallow water/intertidal species of invertebrates were also readily available from year to year, but infaunal benthic species varied considerably in their availability (see Chapter 8).

The following were found to be continuously available, and have been monitored routinely for at least 20 years each:

- Rockweed (*Fucus*)—a marine alga growing on rocks
- Eelgrass (*Zostera*)—a marine grass growing in sediments
- Mussels (*Mytilus edulis*)—a particle feeder growing on rocks and dock pilings in the intertidal zones
- Butter clam (*Saxidomus giganteus*)—a shallow water/intertidal infaunal (sediment-inhabiting) particle feeder
- Littleneck clam (*Protothaca staminea*)—a shallow water/intertidal infaunal particle feeder
- Dungeness crab (*Cancer magister*)—a carnivore/omnivore selected to monitor biomagnification from its feedstock

Figures 10.2–10.4 show cases where there were elevations in trace metals. Mussels (Figure 10.2) showed the greatest elevations in cadmium, copper, and zinc at the ICM dock. There are indications of return to low values from 1997, especially with Cu. Note that because the mussels are from dock pilings, not from the tailings beds, they probably reflect the presence of surface-settled fugitive dust from concentrate loading. The other bivalve mollusks monitored do not show increases.

Figure 10.3 shows increases in copper and zinc in algal (*Fucus*) tissues from Rupert Inlet. There are elevations over time with decreases from 1996 after mine closure. With

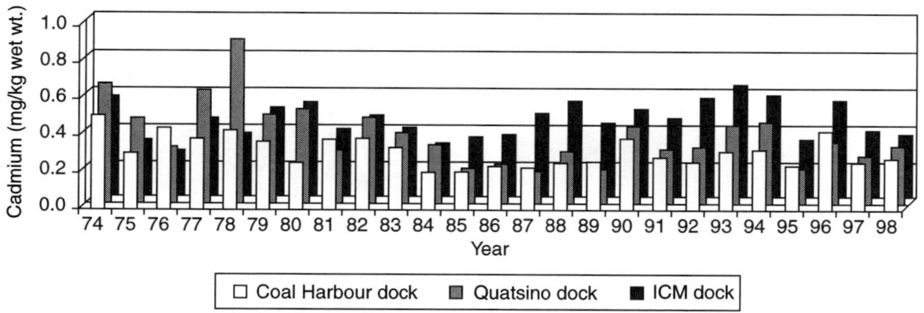

Blue mussel tissue cadmium concentration, 1974–1998

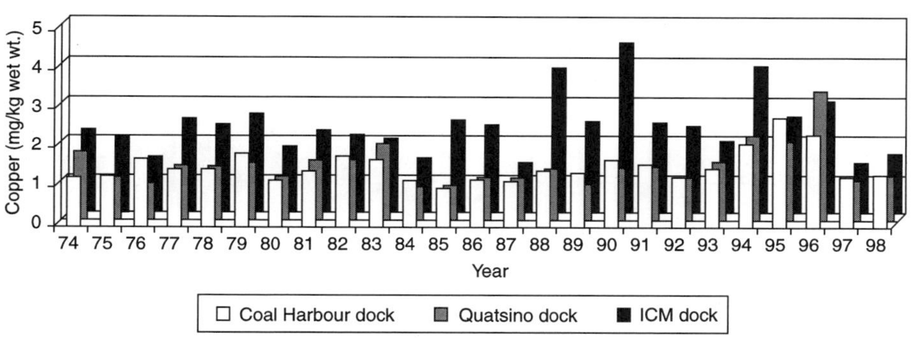

Blue mussel tissue copper concentration, 1974–1998

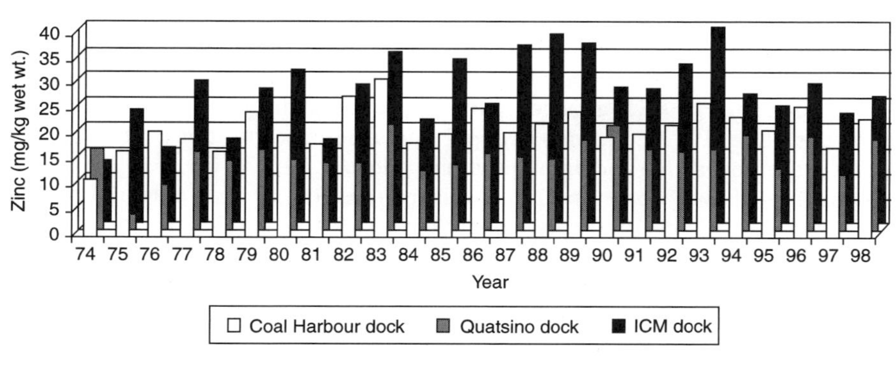

Blue mussel tissue zinc concentration, 1974–1998

FIGURE 10.2 Trace metals in mussels (*Mytilus edulis*)

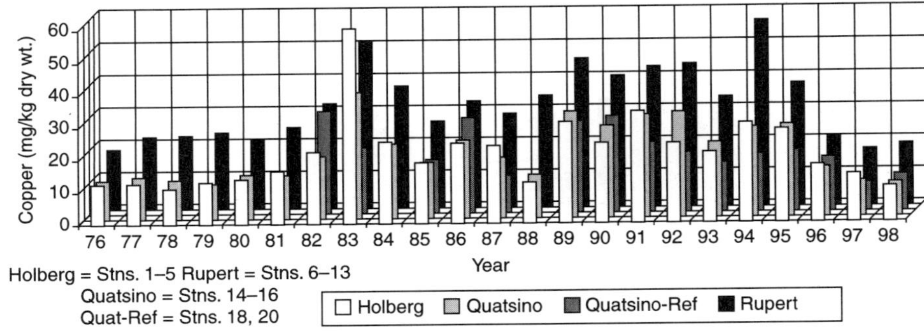

Fucus tissue copper concentration, 1976–1998

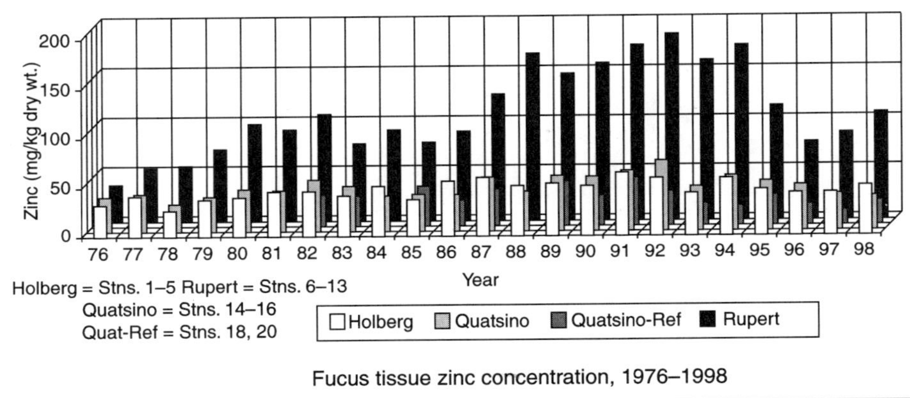

Fucus tissue zinc concentration, 1976–1998

FIGURE 10.3   Trace metals in rockweed (genus *Fucus*)

marine plants, there is some possibility that difficulty in cleaning the surface of attached particles (held there in algae by surface mucus) means that slightly elevated metal levels reflect the presence of such attached particles, rather than absorbed and assimilated trace metals (Ellis 1979). Eelgrass (copper and zinc concentrations) showed similar minor elevations over time and among sampling sites.

Figure 10.4 shows copper and zinc levels in crabs (chosen as a potential biomagnifier through its carnivore/omnivore trophic niche). Test and reference areas fluctuate synchronously, indicating no biomagnification through this potential route.

In conclusion, the mine's trace metal detection systems proved effective in the sense that elevations could be detected (e.g., in mussels on the concentrate loading dock). However, because these mussels are located on the loading dock, the elevations were probably a result of fugitive dust from loading operations, not from the mine tailings discharged at 50 m depth. There were minor elevations of copper and zinc in plants (rockweeds and sea grasses), but these may have been caused by tailings particles attaching to fronds, rather than accumulating within the tissues of the plants. There has been no

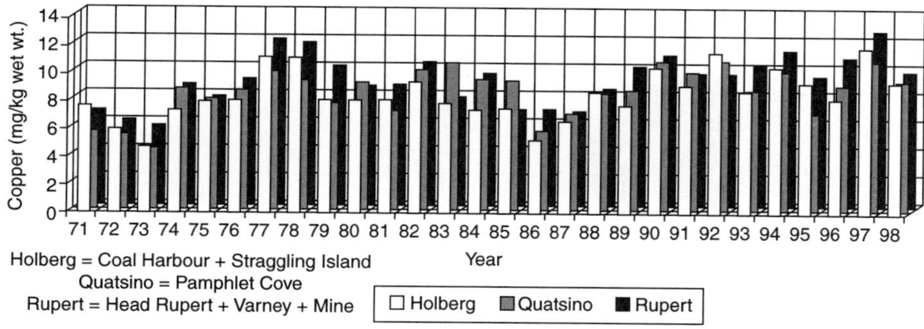

Dungeness crab tissue copper concentration, 1971–1998

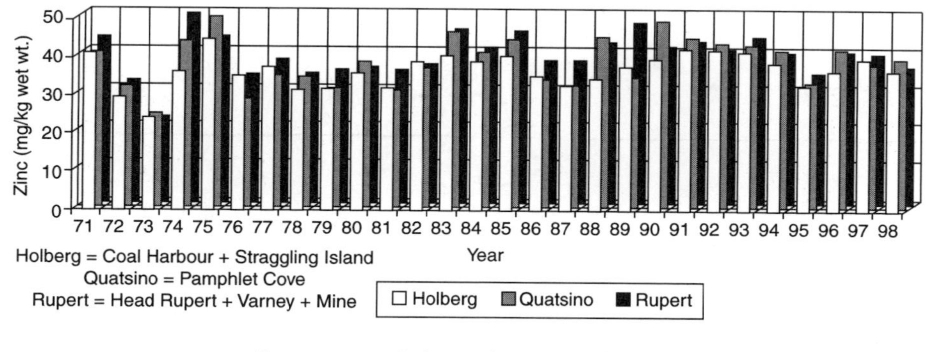

Dungeness crab tissue zinc concentration, 1971–1998

FIGURE 10.4　Trace metals in Dungeness crab (*Cancer magister*)

indication of biomagnification of trace toxins through the most likely food chain in the area, that of benthic invertebrates through Dungeness crab.

These results are not unexpected in view of the insolubility of the copper-bearing rock (chalcopyrite) and the negative results of empirical testing of dissolution of the tailings in situ in the fjord sediments (see previous text).

SETTLEMENT PLATES AND SUSPENDED SOLIDS

Settlement of pelagic larvae onto benthic surfaces has been monitored routinely with a consistent procedure since 1980 by placing settlement plates underwater at the 15 locations shown in Figure 10.5.

Duplicate fiberglass settling plates, 15.2 cm^2, were placed on an anchored support at 0.3-m depth below low tide at each sampling station, three times yearly, and retrieved in March, June, and September each year (except in 1998, with no September samples). The deposited material was dried (105°C), weighed, ashed (550°C), and weighed again. From these weights, total dry solids and the volatile settling rates were calculated. The method is described in detail in ICM (1986).

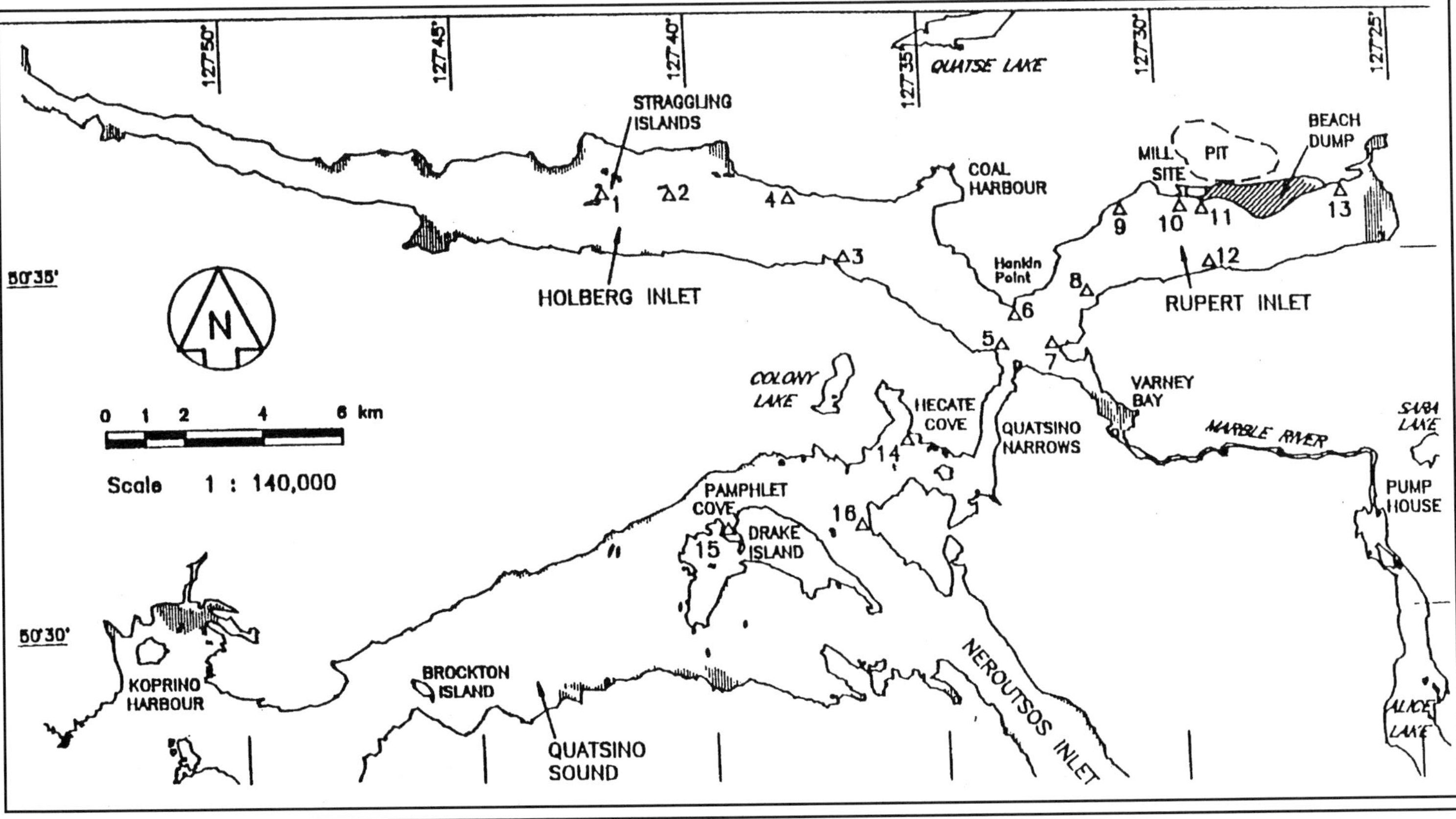

FIGURE 10.5 Map showing sampling stations for settling plates

TABLE 10.3   Settling plate data: settling rates (g/m^2/day) for total and volatile solids

Year	Holberg			Quatsino			Rupert		
	Total	Volatile	% Vol.*	Total	Volatile	% Vol.	Total	Volatile	% Vol.
1980	3.3	0.50	20.2	0.5	0.12	30.0	4.8	0.46	14.5
1981	2.9	0.39	20.9	0.5	0.07	35.9	10.2	0.90	11.7
1982	1.2	0.21	18.4	0.8	0.17	24.7	5.8	0.62	11.8
1983	0.6	0.14	27.3	0.4	0.13	30.2	3.0	0.31	12.8
1984	1.0	0.18	26.9	0.5	0.16	30.4	6.8	0.69	17.9
1985	2.4	0.40	26.9	0.6	0.24	32.1	4.7	0.53	18.0
1986	0.5	0.10	29.4	0.4	0.14	30.1	3.1	0.20	15.2
1987	3.5	0.38	23.5	0.3	0.08	24.1	6.0	0.42	12.3
1988	1.3	0.08	24.2	0.3	0.06	26.3	3.9	0.39	16.0
1989	0.3	0.07	38.7	0.4	0.11	30.4	3.0	0.22	13.9
1990	1.7	0.42	27.2	0.4	0.10	33.9	3.6	0.44	14.2
1991	2.7	0.42	23.3	0.3	0.06	26.1	3.0	0.32	15.0
1992	2.4	0.30	17.8	0.2	0.04	23.9	6.8	0.59	13.0
1993	0.8	0.15	26.8	0.5	0.08	23.3	1.5	0.19	18.5
1994	1.8	0.13	26.9	0.3	0.08	33.5	2.4	0.24	18.8
1995	0.4	0.09	31.4	0.9	0.20	24.8	6.7	0.60	15.7
1996	3.0	0.22	20.6	1.3	0.27	26.2	4.6	0.31	12.0
1997	0.4	0.09	25.5	0.5	0.11	30.5	1.1	0.13	19.9
1998	6.8	0.67	19.5	0.6	0.41	12.9	5.2	0.42	11.8
19-year mean	1.9 ± 1.6	0.26	25.0 ± 5.1	0.50 ± 0.3	0.14	27.9 ± 5.2	4.5 ± 2.2	0.42	14.9 ± 2.7

* % Vol.: Percent volatile.

Table 10.3 shows the results (from ICM 1999). From 1980 to 1996 Rupert Inlet consistently had highest total settlement levels, reflecting the presence of upwelled tailings at station 6, Hankin Point, but the volatile levels were also consistently higher than elsewhere, although a lesser percentage of the total. This annual monitoring presented no evidence that the presence of tailings inhibits levels of colonization by drifting larvae.

A detailed statistical analysis has not been undertaken to confirm these general conclusions.

ACKNOWLEDGMENTS

I am very grateful for cooperation with marine scientists independent of the mine and with a series of mine environmental staff, for the availability of data supporting this chapter, and for the opportunity to discuss data and conclusions based on data analyses. In particular for this chapter, I thank T.F. Pedersen for his analyses of the geochemical data.

REFERENCES

APHA, AWWA, and WPCF. 1971. *Standard Methods for the Examination of Water and Wastewater.* 13th ed. Washington, DC: APHA.

Burd, B.J. 1999. *Post-mining Recovery of Infaunal Benthos in Rupert Inlet.* ICM Annual Environmental Report Vol. III, Part 2. Port Hardy, BC: ICM.

Burd, B.J., and D.V. Ellis. 1994. *ICM Closure Plan: Review of Benthic Surveys 1970 to 1992 for Rupert/Holberg/Quatsino Inlet System.* Port Hardy, BC: ICM.

Ellis, D.V. 1979. *Preliminary SEM Examination of Biological Tissue for Mine Tailings.* Port Hardy, BC: ICM.

Ellis, D.V., and G.C. Jewsbury. 1974. *Turbidity Reviews III. Pacific Salmon–Literature survey and population data.* Report to the UBC Independent Monitoring Group for ICM. Port Hardy, BC: ICM.

Fisheries and Environment Canada. 1977. *Metal Mining Liquid Effluent Regulations and Guidelines.* Report no. EPS-1-WP-77-1. Ottawa, Canada: Fisheries and Environment Canada.

ICM. 1970–1999. Quarterly and Annual Environmental Reports, 1–4 volumes annually. Port Hardy, BC: ICM.

——. 1986. *Methods Manual.* Environment Department. Port Hardy, BC: ICM.

Pedersen, T.F. 1984. Interstitial water metabolite chemistry in a marine mine tailings deposit, Rupert Inlet, B.C. *Canadian Journal Earth Science* 22:1–9.

——. 1985. Early diagenesis of copper and molybdenum in mine tailings and natural sediments in Rupert and Holberg Inlets, British Columbia. *Canadian Journal Earth Science* 22(10):1474–1484.

Pedersen, T.F., and A.J. Losher. 1988. Diagenetic processes in aquatic mine tailings deposits. In *Chemistry and Biology of Solid Wastes.* Edited by W. Salomons and U. Foerstner. Berlin: Springer-Verlag.

Sly, P. 1996. *Review of MEND Studies from 1992 to 1995.* MEND (Mine Environment Neutral Drainage) Report 2.II.Ie.

Postclosure Rehabilitation and Assessment of Inlet System

George W. Poling

During the 25-year operating life of the ICM, while nearly 400 million tonnes of tailing solids were being deposited on the seafloors of Rupert and Holberg Inlets, the regulatory British Columbia Ministry of Environment never once found the mine to be out of compliance with its effluent permits. Granted for the operating life of this mine, the original permit (PE-379) was amended several times (September 21, 1971; December 6, 1971; April 17, 1972; June 30, 1977; December 24, 1985 and June 17, 1993). The amendments were necessary mainly to accommodate increasing production levels. The last amendment was to include permitted limits for both the tailing discharge effluent (up to a maximum of 110,000 m^3/day before seawater admixing) and "water management pond effluent" (up to a maximum annual average rate of 25,000 m^3/day. The latter arose from collection of seepages and runoff from on-land waste storage piles. The most recent permit afforded the ICM was its amended Reclamation Permit (Permit M-9) dated June 8, 1999. Among other things this permit approved the flooding of the open pit, mostly with seawater but topped with a freshwater cap to create a meromictic lake as part of the closure plan.

Figure 11.1 shows an aerial photograph of the ICM site, taken from the west, near the time of mine closure. As part of the closure plan, a channel was cut through from Rupert Inlet to the pit itself, enabling seawater to enter and flood most of the pit's volume. Figure 11.2 shows the seawater fall from the pit side during this flooding process. Approximately 227 million m^3/day of seawater flooded into this pit lake between June 15 and July 23, 1996. This seawater flooded the pit to within about 15 m of sea level before the flooding channel was closed. The top part of the pit lake was next filled with lower density fresh water by pumping in Marble River water and allowing rainwater (at approximately 2 m/yr) to collect at the top. When the water level exceeded sea level by about 2.4 m, the top level of the pit lake water began to overflow from the pit (in early

FIGURE 11.1 Aerial photograph of the ICM site, taken from the west, near the time of mine closure

FIGURE 11.2 The seawater fall viewed from the pit side during the flooding process

FIGURE 11.3 State of land and water reclamation at ICM, June 2001

May 1998). This excess pit lake water exfiltrated through the porous waste-rock beach dump to Rupert Inlet. Figure 11.3 shows an aerial view of the mine site after the pit lake was filled. The artificial pit lake was created to make the freshwater cap, which must meet effluent permit limits, be physically stable, and not be subject to periodic overturns. Metal-contaminated seepage waters emanating from the on-land waste-rock piles could be collected in drainage ditches and injected at depth into the saline layer. Modeling indicated that the saline portion of the pit lake would eventually go anoxic and become a passive treatment system for the acid rock drainages (ARD). Development of sulfate-reducing bacteria (SRB) would generate hydrosulfide ions, which, in turn would precipitate metal contaminants as insoluble metal sulfide particulates. Metal-contaminated ARD waters were injected through two "north" and "south" injector pipes (approximately 1 m in diameter) to a depth of about 220 m below the surface of the pit lake. Beginning in 1997, between 4 and 5 million m^3 of fresh but metal-contaminated waters have been injected each year. This water was injected through diffusers at the terminus of each submerged pipe. This has served to create a three-layer meromictic pit lake. The top brackish layer appears to have stabilized at a thickness of 6–8 m and has a salinity of about 4 parts per thousand. The middle saline layer has a thickness of approximately 217 m and a slowly declining salinity (currently at about 26 parts per thousand). The bottom saline layer extends to approximately 350-m depth and has a salinity of approximately 28 parts per thousand. By mid 2001 the oxygen level of the middle layer had decreased from an initial value of about 10 ppm oxygen to just less than 3 ppm oxygen. In the bottom layer oxygen levels have dropped to 0.03 ppm, and some indication of hydrosulfide generation has been observed at the sediment interface.

Figure 11.3 shows the state of reclamation of the ICM, mid 2001. Most of the buildings and the physical plant have been removed, and the site has been recontoured, till covered, and replanted to restore wildlife habitat. The mine site has won several reclamation awards from the Technical and Research Committee for Reclamation of British Columbia. Because land dumps continue to seep metal-contaminated drainages, work to prevent these seeps from adversely affecting both freshwater and seawater environments continues around this site. Research on adapting the pit lake as a passive ARD treatment system is ongoing. The reclamation bond for this mine is currently at $4 million.

POST-CLOSURE MONITORING OF THE MARINE INLETS

Monitoring of the physical and chemical characteristics of the tailing sediment deposits, the overlying water column, and the inlet biota has continued during the past 6 years (after mine closure), albeit at a somewhat reduced sampling frequency. To understand better the morphological changes from pre-mine to post-mine history, detailed bathymetric and seismic surveys were again conducted in January 2000. As presented in Chapter 4, tailing deposition stayed generally below 50–60-m depth and occupied approximately 11% of the inlet volume. Although some upwelled tailings did escape out of Rupert/Holberg Inlets via Quatsino Narrows, field surveys indicate that the amount that might have escaped was less than 0.4%.

Bathymetric and seismic surveys show that some of the deposited tailings in the vicinity of Hankin Point are being scoured out and redistributed, mostly by seawater density currents from Quatsino Narrows during incoming tides. These tailing sediments are being redeposited mainly up the northern flanks of Rupert Inlet, much like the pre-mine natural sediments were asymmetrically distributed before 1971 when the mine started up. The scour hole developing off Hankin Point has side slopes of approximately 8 degrees and might erode to the original seabed depth of about 170 m by the year 2003. Some redistribution of sediments, then, is still occurring in this inlet system, but most of the sea bottom area appears stable. This is evidenced by the existence of mature benthos communities (see, for example, Chapter 8).

Physical and chemical changes to the seawater columns in both Rupert and Holberg Inlets, caused by the tailing discharges, were either not measurable or barely detectable and were judged not to be harmful in Chapter 6. Salinity, dissolved copper, and arsenic did not change. During the mine's operating life, dissolved manganese increased slightly (less than 2 ppb); dissolved zinc, however, showed a detectable decrease. Post-closure behaviors of these two metals reversed trends with manganese levels dropping (from 5.27 to 2.88 ppb) and zinc levels increasing slightly (2.06 to 4.02 ppb). It was pointed out in Chapter 6 that such slight changes might well have been caused by non-mine-related anthropogenic activities in the watershed. Monitoring has shown conclusively that the euphotic depth in these inlet waters was not affected by the mining activities.

Chapter 7 concluded that the plankton communities in Rupert and Holberg Inlets were not adversely affected by ICM's activities. In addition, tissue metal levels showed no obvious trends that could be attributed to mining activities.

Chapter 8 pointed out that, as originally predicted, the biodiversities of benthos on tailing sediment beds undergoing rapid deposition (greater than 20 cm/yr) were significantly affected during the operational life of this mine. Such areas often showed substantive reductions in the numbers of species present (to approximately 10 species), but only rarely was obliteration found. Upon cessation of high deposition rates, the more opportunistic benthos rapidly recolonized such areas (within about 1 year). Recovery of benthos population densities and diversities was generally complete within 3 years.

Persistence of commercially viable stocks of predatory Dungeness crabs, even during the highest rates of tailing discharges to these inlets, reinforces the high general level of colonization by bottom-dwelling benthos. No biomagnification of heavy metals was detected in these predatory crab populations.

In summary and in retrospect, it now appears that submarine tailing placement of ICM tailings resulted in far less environmental impact than would have accrued had the alternative of on-land disposal been adopted. Although some have concluded that the tailing solids were dispersed more widely than predicted, it now appears that more than 99% of the tailing solids ended up in the deepest regions of Rupert Inlet—almost exactly as was predicted by the mine proponent in 1970. Considering that much has been developed in the past 30 years concerning abilities to predict or model depositional footprints, the agreement between the simplistic prediction of 1970 and the actual depositional footprint is remarkable.

. . . .
Appendix A

Copy of Permit Issued to Utah Construction and Mining Ltd. for the Island Copper Mine on January 20, 1971

ALL COMMUNICATIONS TO:
POLLUTION CONTROL
NT BUILDINGS, VICTORIA, B.C.

AGE REFER TO FILE NO. 0262100-P 379-7
YOUR REF.:

DEPARTMENT OF LANDS, FORESTS, AND WATER RESOURCES
WATER RESOURCES SERVICE

POLLUTION CONTROL BRANCH
VICTORIA, B.C.

<u>REGISTERED</u>   January 20, 1971.

Utah Construction & Mining Co.
Burrard Building,
Suite 1403-1030 West Georgia Street,
VANCOUVER, British Columbia.

Dear Sir:

<u>LETTER OF TRANSMITTAL</u>

Enclosed is a copy of Pollution Control Permit #379-P in the name of
Utah Construction & Mining Co. Your attention is respectfully directed
to the conditions outlined in the Permit.

In conjunction with this Permit you are directed to comply with the following additional requirements.

1.   The Permittee shall engage an independent agent or organization
     to set up and conduct a two-phase sampling and surveillance
     program which will be carried on for at least five (5) operating
     years after discharge commences - all subject to approval by the
     Director.

2.   Prior to discharge the sampling and surveillance program (Phase 1)
     shall incorporate the following objectives which shall be approved
     and initiated in the field by March 1, 1971.

     (a) To ascertain the natural conditions in the receiving
          environment through monitoring the variations in
          physical-chemical-biological characteristics of
          Rupert Inlet, Holberg Inlet, Quatsino Narrows and
          related waters.

     (b) To establish by simulated techniques the vertical and
          horizontal zone and degree of influence of the discharge.

     (c) To establish sampling control stations.

.....2

0262100-P 379-P

2.   January 20, 1971

3.   Upon discharge, the sampling and surveillance program (Phase 2) shall incorporate the following objectives:

(a)   To ascertain the effects of the effluent discharged into the receiving environment through monitoring of the physical-chemical-biological characteristics of Rupert Inlet, Holberg Inlet, Quatsino Narrows and related waters.

(b)   To continue the sampling of control stations.

(c)   To sample and monitor effluent characteristics.

4.   With respect to Items 2 and 3, the following are required:

(a)   Analyse all samples in accordance with the procedures of the most recent issue of "Standard Methods for Examination of Water and Wastewater" published by the American Public Health Association or in any manner approved by the Director.

(b)   Prior to and for the first year after commencing to discharge, data collected during each quarter shall be tabulated and submitted quarterly to the Director.

(c)   Submit comprehensive reports containing the tabulated data of the sampling and surveillance program. The data is to be interpreted and submitted to the Director in such a form that it may be published. The first report shall be submitted approximately at the time of but prior to commencement to discharge and at annual intervals thereafter. This program may be modified at any time at the discretion of the Director.

5.   As a part of the Works the Permittee is required to:

(a)   Install a flow measuring device on the underflow line from the thickeners so that flows may be recorded daily.

(b)   Install a suitable automatic alarm on the effluent disposal system.

(c)   Design and construct an emergency tailings pond to acceptable engineering standards and such shall be maintained in good repair throughout the life of the permit.

3.   January 20, 1971

5.   (c) continued

Note (i)

A record shall be maintained of all the solids and liquids
discharged to the tailings impoundment and this information
shall be made available to the Director.

Note (ii)

Supernatant from this impoundment, if not recycled, is to be
discharged to the inlet at the authorized point of discharge
in accordance with the conditions of this permit.

In conclusion, the Permittee is to take note that prior to implementing
any new change in process that may affect the quality or quantity of
the discharge the Director is to be notified. The Director is also to be
notified of any emergency condition that develops beyond the Permittee's
control and which prevents continued operation of the approved method of
waste disposal.

For your reference enclosed is a copy of the Pollution Control Act, 1967
as amended.

Yours truly,

W.N. Venables, P. Eng.
Director.

AJC/RCS:gMcC
encl.

cc   Health Branch

District Manager, COAST

DEPARTMENT OF LANDS, FORESTS, AND WATER RESOURCES
WATER RESOURCES SERVICE

POLLUTION CONTROL BRANCH

PROVISIONAL PERMIT

Under the Provisions of Section 5 (1) of the Pollution Control Act, 1967

UTAH CONSTRUCTION & MINING CO.
____
(Name.)

of Burrard Building – Suite 1403-1030 West Georgia Street, Vancouver, British Columbia
(Address.)

is hereby authorized to discharge effluent from a copper and molybdenum ore dressing plant
(Plant, factory, municipality, etc.)

located at approximately 9 miles south of Port Hardy subject to the conditions set out below and to the conditions set forth in any appendix attached hereto.

(a) The approximate point of discharge to Rupert Inlet
is located as shown on the attached plan.

(b) The quantity of effluent which may be discharged is to Rupert Inlet from the tailings
thickeners shall not exceed a 24 hour average of 9,300,000 g.p.d.

(c) The characteristics of the effluent described in item (b) above shall be at all times equivalent to or better than that described on
the attached Appendix A Item (c) numbered 1 to 16 inclusive.

(d) The works authorized to be constructed are thickeners, submerged discharge pipeline and
related appurtenances approximately located as shown on the attached plan.

(e) The land from which the effluent originates and to which this permit is appurtenant is that described
on the attached Appendix B

(f) The appendix to which back document contains x nx nx nx nx nx nx nx dealing with
Appendix C attached hereto contains five items
_______________________________________, which are an integral part of this permit.

(g)

APPENDIX "A"

to Pollution Control Permit No. 379-P

ITEM C: The characteristics of the effluent described in item (b) above shall be equivalent to or better than: .

1.	Suspended Solids	500,000 p.p.m.
2.	Total Solids	500,000 p.p.m.
3.	pH range	9.5 to 10.0
4.	cyanide	0.05 p.p.m. (total)
5.	zinc	0.022 p.p.m. (dissolved)
6.	Arsenic	0.008 p.p.m. (dissolved)
7.	Molybdenum	0.02 p.p.m. (dissolved)
8.	Cadmium	0.05 p.p.m. (dissolved)
9.	Chromium	0.05 p.p.m. (dissolved)
10.	Copper	0.005 p.p.m. (dissolved)
11.	Lead	0.005 p.p.m. (dissolved)
12.	Cobalt	0.05 p.p.m. (dissolved)
13.	Nickel	0.005 p.p.m. (dissolved)
14.	Manganese	0.005 p.p.m. (dissolved)
15.	Iron	0.2 p.p.m. (dissolved)
16.	Mercury	0.0003 p.p.m. (total)

APPENDIX "B"

to Pollution Control Permit No.___379-P___

ITEM (c)   The land from which the effluent originates and to which this permit is appurtenant is:

ALL AND SINGULAR that certain parcel or tract of land situate, lying and being portions of Section 24, Township 10 and Lot 15, Rupert District which parcel may be more particularly described as follows:-

COMMENCING at a point distant North 540 feet and West 750 feet of the S.E. corner of said Section 24; thence North 1400 feet; thence West 1300 feet; thence North 600 feet; thence West 2550 feet; thence South 1750 feet, more or less, to an intersection with the southerly boundary of Lot 2155; thence S 70° 13' E and along the said southerly boundary 1050 feet, more or less, to the highwater mark on Rupert Inlet; thence easterly and following the sinuosities of the said highwater mark 5000 feet, more or less, to the point of commencement and containing 219 acres, more or less.

ate___January 20, 1971___

APPENDIX "C"

to Pollution Control Permit No. 379-P

f)   The authority to discharge under this Permit is contingent upon the works authorized having been constructed as per final construction plans approved in accordance with the Pollution Control Act, 1967.

g)   All the works including the emergency spill basin are to be constructed on or before December 31, 1973 or prior to discharge whichever date is the sooner.

h)   The Permittee shall not fail to secure tenure to land sufficient and suitable for a tailings disposal pond on or before December 31, 1973, or prior to discharge, whichever date is the sooner.

i)   The Permittee shall not fail to establish and carry out a "Two-Phase" sampling and surveillance program(s) acceptable to the Director. The sampling and surveillance program under Phase 1 shall be initiated prior to discharge and under Phase 2 subsequent to discharge. Phase 2 of this program shall continue for a minimum period of five (5) operating years after commencement to discharge.

j)   The Permittee shall not fail, prior to commencement to discharge, to post security in an acceptable form in the amount of $1,500,000.00 for a period of five (5) operating years after discharge commences. The security or a portion thereof will be subject to forfeiture should the Permittee fail to comply with an order of the Director to construct an alternate or modified system of treatment and/or discharge.

Date   January 20, 1971

Director of Pollution Control.

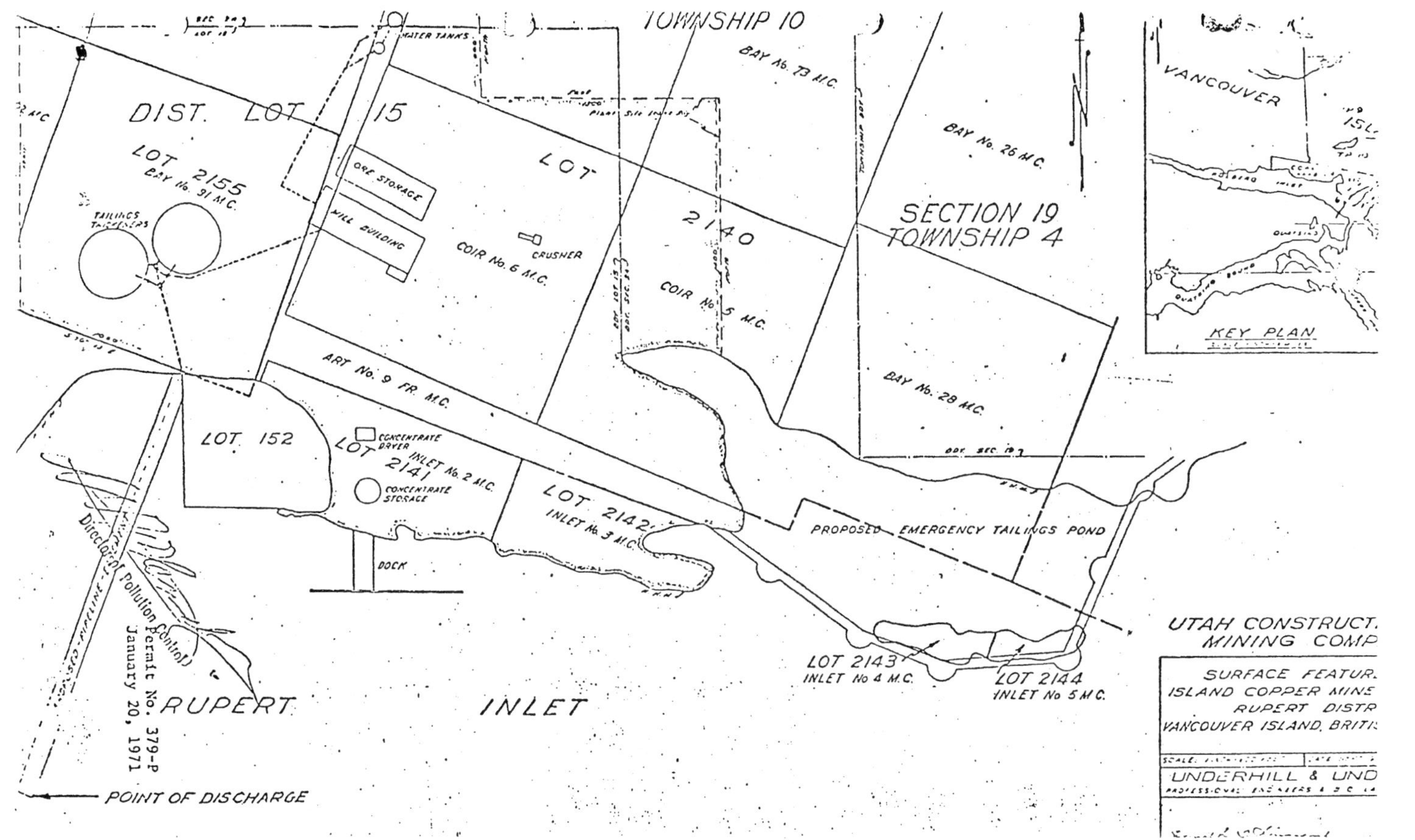

TOWNSHIP 10
DIST. LOT 15
LOT 2155
BAY No. 91 M.C.
TAILINGS THICKENERS
ORE STORAGE
MILL BUILDING
WATER TANKS
LOT
COIR No. 6 M.C.
CRUSHER
BAY No. 73 M.C.
BAY No. 26 M.C.
SECTION 19
TOWNSHIP 4
2140
COIR No. 5 M.C.
VANCOUVER
KEY PLAN
ART No. 9 FR. M.C.
LOT 152
CONCENTRATE DRYER
LOT 2141
INLET No. 2 M.C.
CONCENTRATE STORAGE
DOCK
LOT 2142
INLET No. 3 M.C.
BAY No. 28 M.C.
PROPOSED EMERGENCY TAILINGS POND
Director of Pollution Control
Permit No. 379-P
January 20, 1971
RUPERT
INLET
LOT 2143
INLET No 4 M.C.
LOT 2144
INLET No 5 M.C.
POINT OF DISCHARGE
UTAH CONSTRUCT.
MINING COMP.
SURFACE FEATUR.
ISLAND COPPER MINE
RUPERT DISTR.
VANCOUVER ISLAND, BRITI.
UNDERHILL & UND.

. . . .
Appendix B

Copy of the Latest Permit Covering the Island Copper Mine's Production Period

PROVINCE OF
BRITISH COLUMBIA

ENVIRONMENTAL PROTECTION

2569 Kenworth Road
Nanaimo, British Columbia
V9T 4P7
Telephone:(604) 751-3100

MINISTRY OF ENVIRONMENT,
LANDS AND PARKS

PERMIT

Under the Provisions of the Waste Management Act

BHP Minerals Canada Ltd.

2800 Park Place

666 Burrard Street

Vancouver, British Columbia

V6C 2Z7

is authorized to discharge effluent to the land from a water
management pond and to Rupert Inlet from a copper-molybdenum
mine/mill complex located near Port Hardy, British Columbia subject
to the conditions listed below. Contravention of any of these
conditions is a violation of the Waste Management Act and may result
in prosecution.

1. **SPECIFIC AUTHORIZED DISCHARGES AND RELATED REQUIREMENTS**

   1.1 Mill Tailings Effluent

   1.1.1 The maximum rate at which effluent may be discharged,
   prior to seawater mixing, is 110 100 m³/day with an annual
   average not to exceed 95 000 m³/day.

G.E. Oldham, P. Eng.
Regional Waste Manager

Date issued: January 20, 1971

Date amended: JUN 1 7 1993
(most recent)

Page: 1 of 11

Permit No. PE-379

1.1.2 The characteristics of the effluent, prior to mixing
with seawater, shall be equivalent to or better than:

Total Suspended Solids	500 000 mg/l
pH	7.5 - 11.5
Dissolved Cyanide	0.10 mg/l
Dissolved Arsenic	0.10 mg/l
Dissolved Molybdenum	1.0 mg/l
Dissolved Cadmium	0.005 mg/l
Dissolved Copper	0.05 mg/l
Dissolved Lead	0.05 mg/l
Toxicity (96 hour LC_{50})*	100% effluent concentration

* analyzed on a decanted sample.

1.1.3 The works authorized are thickeners (optional as
required to meet the conditions of Section 1.1.1, Section 1.1.2
and if approved by the Regional Waste Manager under Section
11), submerged outfall system, standby submerged outfall
system, tailings alarm system, pipelines, and related
appurtenances approximately located as shown on the attached
Appendix A-1.

1.1.4 The works authorized must be complete and in operation
on and from the date of this amended permit.

1.2 Water Management Pond Effluent

1.2.1 The maximum rate at which effluent may be discharged is
an annual average of 25 000 m³/day.

1.2.2 The characteristics of the effluent shall be equivalent
to or better than:

pH	6.5 - 11.5
Dissolved Arsenic	0.10 mg/l
Dissolved Molybdenum	0.50 mg/l
Dissolved Cadmium	0.01 mg/l
Dissolved Copper	0.05 mg/l
Dissolved Lead	0.05 mg/l
Dissolved Zinc	1.0 mg/l
Toxicity (96 hour LC_{50})	100% effluent concentration

G.E. Oldham, P. Eng.
Regional Waste Manager

Date issued: January 20, 1971

Date amended: JUN 1 7 1993
(most recent)

Page: 2 of 11

Permit No. PE-379

1.2.3   The works authorized are a lined water management pond,
an exfiltration pond, and related appurtenances approximately
located as shown on the attached Appendix A-1.

1.2.4   The works authorized must be complete and in operation
on and from the date of this amended permit.

2.   **LOCATION OF THE FACILITIES**

Lot 15, Section 24, Township 10, Rupert Land District.

3.   **TAILINGS EFFLUENT SYSTEMS**

The Permittee is to maintain two tailings effluent systems in
satisfactory working order.

4.   **POSTING OF SECURITY**

The Permittee shall maintain security with the Minister of
Finance and Corporate Relations, as a condition of the
Reclamation Permit issued by the Ministry of Energy, Mines and
Petroleum Resources.

5.   **MILLING REAGENTS**

The Permittee is to inform the Regional Waste Manager prior to
changing the reagents used in the milling process, except for
those used on a short-term experimental basis. At the end of
each calendar year, the Permittee is to advise the Regional
Waste Manager of the amount and period of usage of each reagent
used during the preceding year.

6.   **PROCESS MODIFICATIONS**

The Permittee shall notify the Regional Waste Manager prior to
implementing changes to any process that may affect the quality
and/or quantity of the discharges.

G.E. Oldham, P. Eng.
Regional Waste Manager

Date issued: January 20, 1971

Date amended: JUN 17 1993
(most recent)

Permit No. PE-379

7.   <u>MAINTENANCE OF WORKS</u>

The Permittee shall inspect the pollution control works
regularly and maintain them in good working order. Notify the
Regional Waste Manager of any malfunction of these works.

8.   <u>EMERGENCY PROCEDURES</u>

In the event of an emergency or condition beyond the control of
the Permittee which prevents continuing operation of the
approved method of pollution control, the Permittee shall
immediately notify the Regional Waste Manager and take
appropriate remedial action.

9.   <u>BYPASSES</u>

The discharge of effluent which has bypassed the authorized
works is prohibited unless the approval of the Regional Waste
Manager is obtained and confirmed in writing.

10.   <u>CONTINGENCY PLAN</u>

The Permittee shall prepare and submit a contingency plan to
the Regional Waste Manager, on or before September 30, 1993,
which outlines the management options available for the
discharge from the water management pond to the exfiltration
pond to ensure that the criteria specified in Section 1.2.2 are
not exceeded.

11.   <u>DISCHARGE OF UNTHICKENED TAILINGS</u>

The discharge of unthickened tailings into Rupert Inlet is
subject to the written approval of the Regional Waste Manager.

G.E. Oldham, P. Eng.
Regional Waste Manager

Date issued: January 20, 1971

Date amended: JUN 1 7 1993
(most recent)

Page: 4 of 11

Permit No. PE-379

12. MONITORING

The Permittee is required to undertake the following monitoring program. The minesite and receiving environment sampling locations are designated in accordance with established stations as defined in the 1991 Annual Environmental Assessment Report.

12.1 Mill Tailings Effluent

	Station	Frequency	Analysis
(a) Physical	•prior to seawater dilution	•daily	•flow (m^3/day) •solids discharge (tonnes/day)
(b) Chemical	•prior to seawater dilution	•weekly composite of daily grab samples (2/day, 7 days/week)	•% solids, pH, As(d), Cd(d), Cu(d), Pb(d), Mo(d), Zn(d), Mn(d)
	•prior to seawater dilution	•monthly grab sample to be taken when toxicity sample is taken *	•CN(d) if toxicity <100%
(c) Biological	•prior to seawater dilution	•monthly grab sample	•toxicity (96hr LC_{50}) on a decanted sample.

* Specific preservation and storage methods will be issued by the Regional Waste Manager.

G.E. Oldham, P. Eng.
Regional Waste Manager

Date issued: January 20, 1971

Date amended: JUN 1 7 1993
(most recent)

Permit No. PE-379

12.2 Offsite Drainages

	Station	Frequency	Analysis
(a) Chemical	•Frances Lake •Twin Lakes	•quarterly grab samples (March, June, September, December)	•dissolved oxygen, pH, total suspended solids, hardness, conductivity, turbidity, total alkalinity/acidity, Ca(t), SO_4(t), Cd(d), Cu(t+d), Fe(d), Pb(d), Mn(d), Mo(d), Zn(t+d), Mg(t)

12.3 Minesite Drainage

	Station	Frequency	Analysis
(a) Physical	•Water Management Pond Outflow to Exfiltration Pond	•weekly	•daily average flow (m^3/day)
(b) Chemical	•East Drainage Ditch •1080 Ditch •Pit Water •Water Management Pond	•monthly grab samples	•pH, hardness, conductivity, total alkalinity/acidity, Ca(t), SO_4(t), As(d), Cd(d), Cu(d), Fe(d), Pb(d), Mn(d), Mo(d), Zn(d), Mg(t)
	•Water Management Pond	•weekly grab samples	•Cu(t), Zn(t)

G.E. Oldham, P. Eng.
Regional Waste Manager

Date issued: January 20, 1971

Date amended:
(most recent) JUN 1 7 1993

Page: 6 of 11

Permit No. PE-379

12.4 Marine (Physical)

	Station	Frequency	Analysis
(a) Tailings Distribution			
(i) Marine Sediment Coring	•24 stations	•annually (March)	•visual to determine distribution of tailings less than 60 cm thick. Five station cores to be analyzed for copper and molybdenum
(ii) Marine Sediment Analysis	•24 stations (to include stations 2, 6, 11, 19, 23, 24, 28, 30)	•annual bottom surface grab samples (March)	•copper and molybdenum
(iii) Echo Sounding Survey	•18 stations	•annually (March)	•echogram
(b) Suspended Sediment Distribution			
(i) Water Sampling	•4 stations (A, B, D, E) at standard depths of 0, 5, 10, 20, 30, 45, 75, 100, 150, and 10 metres from bottom.	•quarterly grab profiles (March, June, September, December)	•NTU and gravimetric analysis
(ii) Suspended Sediment Settling Traps	•4 stations (2 depths per station)	•quarterly grab profiles (March, June, September, December)	•total and fixed solids

G.E. Oldham, P. Eng.
Regional Waste Manager

Date issued: January 20, 1971

Date amended: JUN 17 1993
(most recent)

Page: 7 of 11

Permit No. PE-379

12.5 Marine (Chemical)

	Station	Frequency	Analysis
(a) Chemical	•5 stations (A, B, C, D, E) at 4 depths (0, 5, 30, and 10 metres from the bottom)	•quarterly grab profiles (March, June, September, December)	•dissolved oxygen, Cu(d), Mn(d), Mo(d), Zn(d)
	•3 stations (X, Y, Z) at 3 bottom depths (5m, 30m and toe of beach dump)	•bottom surface grab samples (January, March, April, May, June, September)	Cu(d), Zn(d), Cd(d)

12.6 Marine (Biological - Plants)

	Station	Frequency	Analysis
(a) Phytoplankton	•7 stations, 4 depths	•grab profiles 3 times annually (May, June, September)	•chlorophyll "a"
(b) Periphyton	•16 stations	•quarterly settling plate analysis (March, May, July, September)	•total and volatile solids

G.E. Oldham, P. Eng.
Regional Waste Manager

Date issued: January 20, 1971

Date amended: JUN 1 7 1993
(most recent)

Page: 8 of 11

Permit No. PE-379

PROVINCE OF
BRITISH COLUMBIA

ENVIRONMENTAL PROTECTION

12.7 Marine (Biological - Animals)

	Station	Frequency	Analysis
(a) Zooplankton	(i) 4 stations, horizontal tows at 3 depths plus 1 vertical haul.	•semi-annual tows (March, September)	•detailed identification
	(ii) 4 stations using night tows.	•semi-annual tows (March, September)	•metals analysis (Cu, Zn, As, Pb, Cd) of zooplankton on whole sample and sorted for euphausiids
(b) Benthic organisms	•24 stations (3 samples each)	•annual grab samples (September)	•identify infauna from sediment samples
(c) Crabs	•6 stations	•annual trapping (June)	•weigh, measure, identify sex, and analyze for Cu, Pb, Zn, As, and Cd •analyze for Hg during final year of milling operations
(d) Clams	•6 stations •collect marker species of clams and other bivalves if available	•annual collection (June)	•weigh, measure, and analyze for Cu, Pb, Zn, As, and Cd
(e) Mussels	•5 stations; BHP dock, Coal Harbour dock, Quatsino dock, Port Hardy and Winter Harbour	•annual collection (June)	•weigh, measure, and analyze for Cu, Pb, Zn, As, and Cd

G.E. Oldham, P. Eng.
Regional Waste Manager

Date issued: January 20, 1971

Date amended: **JUN 17 1993**
(most recent)

Page: 9 of 11

Permit No. PE-379

13. <u>ANALYSES</u>

Analyses are to be carried out in accordance with procedures
described in "A Laboratory Manual for the Chemical Analysis of
Waters, Wastewaters, Sediments and Biological Materials, (1976
edition including updates)", April 1989, or by suitable
alternative procedures as authorized by the Regional Waste
Manager.

Analyses for determining the toxicity of liquid effluents to
fish shall be carried out in accordance with the procedures
described in the "Provincial Guidelines and Laboratory
Procedures for Measuring Acute Lethal Toxicity of Liquid
Effluents to Fish" November 1982. The Regional Waste Manager
will advise the Permittee which method of measurement for
expressing lethal toxicity shall be used. The method of
sampling and the method of bioassay will be determined by the
Regional Waste Manager.

Copies of the above manuals are available from the
Environmental Protection Division, Ministry of Environment,
Lands and Parks, 777 Broughton Street, Victoria, British
Columbia, V8V 1X4, at a cost of $70.00 and $5.00 respectively
and are also available for inspection at all Environmental
Protection offices.

It should be noted that analysis for toxicity (Section 1.1.2)
is to be carried out on decanted effluent samples. Established
procedure involves letting the samples settle for 18 to 24
hours before decanting.

14. <u>REPORTING</u>

The monitoring program shall be carried out by the Permittee
and the results submitted to the Regional Waste Manager as
follows:

14.1 All effluent data for the discharge authorized by Section
1.1 including the flow measurements are to be submitted
monthly.

G.E. Oldham, P. Eng.
Regional Waste Manager

Date issued: January 20, 1971

Date amended: JUN 17 1993
(most recent)

Page: 10 of 11

Permit No. PE-379

14.2 The annual report, including all the monitoring results
for the period January 1 to December 31, is to be
submitted on or before September 30 of each year. The
report should consist of the data plus a summary prepared
by BHP Minerals Canada Ltd. under the guidance of the
Environmental Advisors.

Based on the results of the monitoring program, the Permittee
monitoring requirements may be extended or altered by the
Regional Waste Manager.

15. **SPILL REPORTING**

All spills to the environment (as defined in the Spill
Reporting Regulation) shall be reported immediately in
accordance with the Spill Reporting Regulation. Notification
shall be via the Provincial Emergency Program at
1-800-663-3456.

16. **ENVIRONMENTAL ADVISORS - REVIEW**

Prior to submission of the annual report, the Environmental
Advisors consisting of representation from the fields of marine
biology, oceanography, geology and mineral engineering shall
critically review the document. Any changes to the
representation on the Environmental Advisory group shall be
approved by the Regional Waste Manager. Representatives of the
advisory group should visit the site as required to review the
field and laboratory procedures and provide guidance to the
Permittee.

G.E. Oldham, P. Eng.
Regional Waste Manager

Date issued: January 20, 1971

Date amended: JUN 17 1993
(most recent)

Permit No. PE-379

Province of
British Columbia

Ministry of
Environment
Lands and Parks

ENVIRONMENTAL PROTECTION

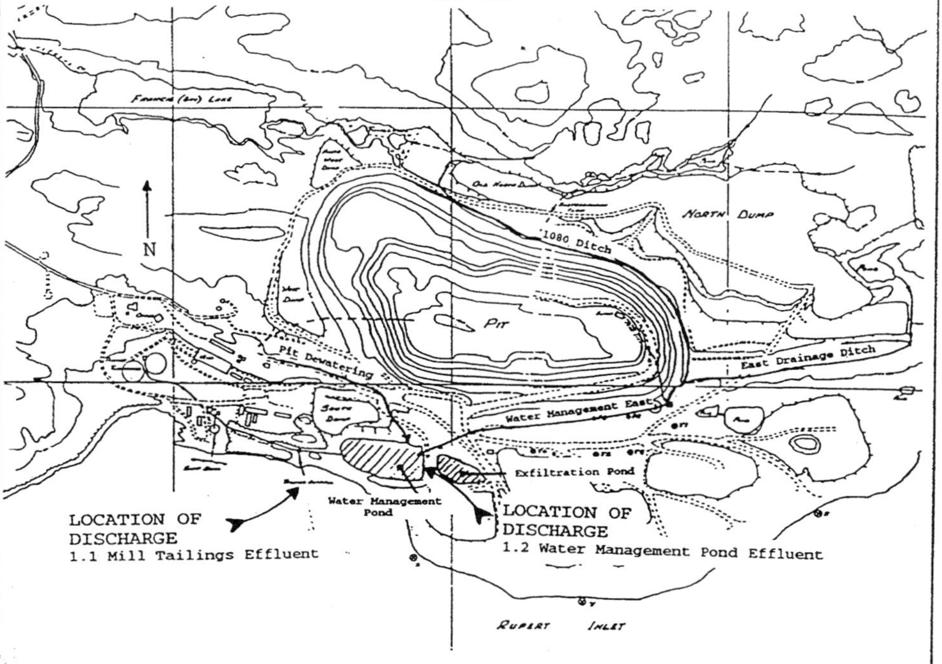

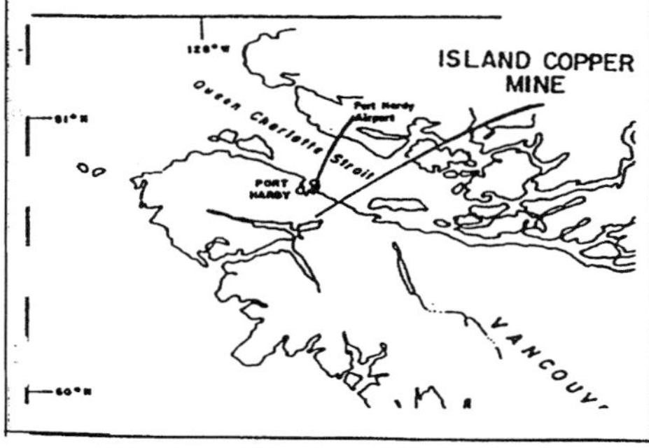

BHP Minerals Canada Limited

Name of Permittee

Appendix ___A-1___ to Permit No. ___PE-379___

Approval No. ___________

Date Issued ___January 20, 1971___

Date Amended ___JUN 17 1993___

Regional Waste Manager

Index

Note: *f* indicates figure; *t* indicates table.

A

Acid generation   8–9, 86
Acid Mine Drainage Study of the North Dump, Island Copper Mine   77
Acidic rock drainage   11
   and pit lake   163
A-J Mine (Alaska)   5
Alaska   3–5
Alaska Bureau of Mines   7
Algae   144
   forests   139–140, 142
   at Hankin Point   197*f*, 198*f*, 199*f*
   increases in copper and zinc   154–156, 155*f*, 156
   at mine dock, 200*f*
   monitoring   154
Alkalinity   106–110, 108*t*
Ammotrypane aulogaster   129
Anemones   141
   at Hankin Point   197*f*, 199*f*
   at mine dock, 200*f*
Annual environmental assessment reports (AEARs)   26, 27
ANOVA   94–95, 106
Arsenic
   correlations with zooplankton   110, 111*t*
   monitoring   154
   neutralization and precipitation in marine sediments   153–154
   post-closure   164
   seawater levels 1971–1992   99–100, 100*t*
   in tailings   90
Asset Surveyor software   49
Axinopsida serricata   129

B

Ball mills   38
Barnacles   144, 145
Batu Hijau (Indonesia)   5
BC Research   23, 86, 90
   tailings bioassays   152
Beach dump   143–145
Benson Lake Mine   17
Benthos   3, 7–8. *See also* Intertidal biodiversity, Rocky shore biodiversity, Seabed biodiversity, Underwater biodiversity
   archiving of sample   119
   infauna samples   117*f*

   infauna surveys   116–120, 154
   infaunal sampling equipment   116, 119*f*
   levels of copper and tailings deposition tolerable to infaunal species   127–128
   monitoring   115–116
   post-closure   164–165
   sustainable ecological succession   132
   tissue monitoring for trace metal bioaccumulation   153–157, 155*f*, 156*f*, 157*f*
BHP Billiton   2, 27
Biodiversity. *See* Intertidal biodiversity, Rocky shore biodiversity, Seabed biodiversity, Underwater biodiversity
Biological effects   3, 7–9, 11–12
Biological properties   105–106, 113
   alkalinity   106–110, 108*t*
   changes after mine closure   112–113, 113*t*
   chlorophyll A   106–110, 108*t*, 109*f*
   field sampling   106
   oxygen   106–110, 108*t*
   pH   106–110, 108*t*
   sampling stations   106, 107*f*
   statistical analyses   106
   total zooplankton   110–112, 110*t*, 111*t*, 112*t*
Biotite-chlorite   85
Black Angel Mine (Greenland)   8–9
Bougainville Copper Mine (Papua, New Guinea)   2
Bray-Curtis coefficient   131
British Columbia, University of   25–26
British Columbia Ministry of Environment   161
Brittle stars   130
Burrowing anemones   141
Burrowing worms   21
Butter clam   154

C

Cadmium
   monitoring   154
   in mussels   154, 155*t*, 156
Calcite   85
Cancer magister. See Dungeness crab
Cancer productus   148
Capitella capitata   129
Carter, Ralf   23
Case Studies of Submarine Tailings Disposal: Volume I—North American Examples   5

Case Studies of Submarine Tailings Disposal: Volume II–Worldwide Case Histories and Screening Criteria 5
Catastrophic failure 9–11
Chaetozone acuta 129
Chalcopyrite 17–19, 35, 88, 153–154, 157. *See also* Copper
    flotation 39–43
Chinook salmon 151, 152
Chlorophyll A 106–110, 108*t*, 109*f*
    before and after mining 112–113, 113*t*
    correlations with zooplankton 110, 111*t*
Chum salmon 151
Clams 131, 132
    monitoring 154
Climax community 128–129
Clinocardium 132
Coagulation 7, 11
Coal Harbour 58, 76
    tailings deposited beyond 79
Coal mining 17
Cockles 145
Coho salmon 151, 152
Compsomyax 132
Copper 85–86
    in algae 154–156, 155*f*, 156
    correlations with zooplankton 110, 111*t*
    dissolved in seawater, 91, 92*f*
    dissolved in tailings 90
    dissolved levels following mine closure 100–102, 102*t*, 103
    elevated levels and biodiversity decline 154
    levels in Dungeness crab 156, 157*f*
    monitoring 154
    in mussels 154, 155*t*, 156
    neutralization and precipitation in marine sediments 153–154, 157
    post-closure 164
    seawater levels 1971–1992 99–100, 100*t*
Cossura pygodactylata 129
Crab 141. *See also* Dungeness crab
Crushing and grinding 38
Crustacea 130–131, 132, 144
    on natural sediment 203*f*
Cuajone Mine (Peru) 2
Cyclone classifiers 38

D

Deep sea tailing placement 2
    advantages 9–11
    as best environmental choice at ICM 24
    biological concerns 7–9
    compared with on-land impoundment 9, 10*t*
    disadvantages 11–12
    evaluation procedure 13–14
    history 2–3
    key publications 3, 5
    and mine closure 11
    mix tanks 6, 6*f*
    operations using or considering DSTP 3, 4*t*
    and proximity to marine coast 2, 5
    and pumping 6
    and regulatory sensitivity 14

    requirements for 2, 9
    risks 2
    slurry pipelines 5–6
    structures and strength of 7, 12
    success of 165
    system schematic 6*f*
    technology 5–7
Density currents 6, 7
Department of Fisheries 19
DSTP. *See* Deep sea tailing placement
Dungeness crab 130, 132
    catch data 148, 150*t*
    copper and zinc levels 156, 157, 157*f*
    data archives 148
    and government statistics 147
    at Hankin Point, 199*f*
    maintenance of 147, 165
    monitoring 148–151, 154
    sampling stations 148, 149*f*
    and trace element biomagnification 147, 148, 150–151

E

Echinoderms 130
Ecosystem effects. *See* Biological effects
Eelgrass 141
    monitoring 154
Eelpouts 145
EG&G Boomer 49
Ellis, D.V. 25, 26
End Creek Fault 35
Environmental Protection Agency. *See* U.S. Environmental Protection Agency
EPC 9800 line scan recorder 51
Epifauna 142, 144
Equilibrium concept 128–129
Euchone incolor 129

F

Feldspar 85
Fish 131
    at Hankin Point 198*f*
    monitoring 154
    on natural sediment 203*f*
    on tailings outfall 202*f*
Fisheries and Environment Canada 152
Flotation 39–43
Flowsheet 38, 39*f*
Foreman, R. 139–140
Fort Rupert 17
Fucus 144, 154–156, 155*f*

G

Galena 88
Gracilaria 140
Greenpeace 24
Grinding 38

H

Halcampa decemtentaculata 130
Hankin Point 47, 58–59, 60
    algae, anemones, mollusks, and sediment 197*f*
    post-closure monitoring 164

sediment loading   140–141
slime-worm and fish   198*f*
tailings upwelling   139, 159
turbidity problems at scour hole   66, 69*f*, 70*f*
Holberg Inlet   2–3, 19. *See also* Morphological
      change (Rupert and Holberg Inlets)
   area maps   18*f*, 35*f*, 46*f*
   biological and chemical characterization   21
   changes in metal levels 1971–1992   99–100, 100*t*
   changes in physical and chemical properties
      93–94
   dissolved metals following mine closure
      100–102, 102*t*, 103
   early lack of tailings infiltration   140
   field sampling   94
   geographical description   45
   mean oceanographic parameters before and after
      mining   95
   near-surface turbidity   95–99, 96*f*, 97*t*, 99*t*, 102
   oceanographic statistical analyses   94–95
   physical characterization   21–23, 22*f*
   post-closure monitoring   164–165
   projected effects of tailing discharge   23–24
   salinity   95, 102
   seawater sampling stations   91, 91*f*, 94, 94*f*
   Secchi depth   95–99, 96*f*, 98*t*, 99*t*
   sediment troughs   116, 118*f*
   tailings deposited beyond Coal Harbour   79
   temperature   95, 102
Holothuria   131
Horne, I.   24

I

ICM. *See* Island Copper Mine
Impoundments
   dusting   2
   on-land compared with DSTP   9, 10*t*
   structural failure   2
   water contamination   2
Intertidal biodiversity, 143
Invertebrates   131
   monitoring   154
Iron oxides   89–90
Island Copper Mine   2
   aerial photograph   161, 162*f*
   area maps   18*f*, 35*f*
   background   2–3, 17–19
   capital costs   43
   compliance with permits   161
   DSTP as best environmental choice   24
   environmental monitoring program and
      reports   27–30, 28*t*–29*t*
   estimated tailings discharge   79
   independent monitoring agency   25–27
   operating costs   43
   permits for DSTP   24–25, 27, 30, 168*f*–175*f*,
      178*f*–189*f*
   pit flooding   161–163, 162*f*, 163*f*
   preoperational tailings studies   19–20, 20*t*, 21*t*
   public hearing on DSTP   24–27
   reclamation   163–164, 163*f*
   Reclamation Permit   161
   and salmon hatchery   151

seawater fall   161, 162*f*
selection of DSTP for tailings disposal   19
technical environmental data sources   27
Isopods   145

J

Johnson, J.W.   23
Jordan River Mine (British Columbia)   2

K

Keefe-Bergeson index   131
Kelp   143, 144
Kensington Gold Mine (Alaska)   5, 8
Ker, Priestman and Associates Ltd.   26
Krohnhite filter set   49

L

LC$_{50}$ 96-hour exposure tests   152–153
Lichen   144
Limpets   144
Littleneck clam   154
Littorina   144
Los Angeles County Museum of Natural
      History   119
Lowermost infra-littoral fringe   144
Lumbrineris luti   129

M

Macoma spp.   130
Macrophytes   140
Magnetite   85
Manganese
   correlations with zooplankton   110, 111*t*
   dissolved in seawater, 91, 92*f*
   dissolved levels following mine closure
      100–102, 102*t*, 103
   oxides   89–90
   post-closure   164
   seawater levels 1971–1992   99–100, 100*t*
Marble River   21, 66
   salmon hatchery   151–152
Marcopper Mine (Philippines)   2
Margalef index   131
Marine Georesources and Geotechnology   5
Marine grasses   154
   copper and zinc elevations   156
Middle littoral zone   144
Milbourne, Gordon   17
Mine dock   200*f*
Mineral processing
   crushing and grinding   38
   cyclone classifiers   38
   flowsheet   38, 39*f*
   froth flotation   39–43
   particle size distribution of ground ore and
      tailings   39, 40*f*
   reagents   42–43, 42*t*
Mini-Fix Range-Range System   51
Mining technique   36, 38*f*
Misty Fjords National Monument   5
Mix tanks   30–32, 31*f*, 32*f*
Mollusks   132
   bivalve   129, 130, 132

at Hankin Point   197f
Molybdenum   17–19, 35, 85
   dissolved in tailings   90
   flotation   39–43
   monitoring   154
Moore, Patrick   24
Morphological change (Rupert and Holberg
      Inlets)   45–47, 46f, 81–82
   bathymetric change   58–59
   bathymetric surveys   49, 51t
   cross-sectional profiles   49, 53f
   data processing and map preparation   51–58
   deposition compared with estimates   80
   deposition of mine tailings   60–61, 62f
   dry density of waste materials   77–78, 78f, 80–81
   echo sounder   49
   estimated deposition in Rupert Inlet   79
   estimated tailings discharge   79
   Global Positioning System   49
   hypsometry curves   71–76, 71f, 72f, 73f, 74f, 75f,
      76f
   loss through Quatsino Narrows   79
   loss to Holberg Inlet   79
   mine material deposition budget   76, 77t, 80t
   post-mine bathymetry   50f
   pre-mine bathymetry   48f
   pre-mine deposition patterns   59–60
   predicted tailings placement   59
   qualitative and quantitative assessments   61–62
   schematic depositional model   61, 62f
   seismic lines   51, 55f
   seismic profile along line 19   51, 56f, 57f
   seismic surveys   49–51, 59
   sounding profiles   49, 54f
   sounding tracks   49, 52f
   turbidity problems at Hankin Point scour
      hole   66, 69f, 70f
   uncertainty of estimates   79–80
   visual change (isopach maps)   62, 67f, 68f
   visual change (superimposed cross-sectional
      profiles)   62, 64f, 65f, 66f
   visual change (superimposed isobaths)   62, 63f
   waste rock in depths below datum   78
Murray, J.   25, 26
Mussels   143, 144
   elevations of cadmium, copper, and zinc   154,
      155t, 156
   monitoring   154
Mytilus edulis   154

N

Neogardiella   140
Nephtys cornuta   129
Neroutsos Inlet   17
Niskin sampler   94
Nitrate   89
North Vancouver Island Salmon Enhancement
      Association   151

O

Odom Echotrak echo sounder   49
Oncorhynchus   148, 151
Oncorhynchus gorbuscha   148, 151
Oncorhynchus keta   151

Oncorhynchus kisutch   151, 152
Oncorhynchus mykiss   152
Oncorhynchus nerka   151
Oncorhynchus tshawytscha   151, 152
Open pit mining   35–36, 37f
Ophelina acuminata   129
Orebody
   characterization   35
   sampling stations   36f
Outfall system   31–32, 31f
Owenia fusiformis   130
Oxidation   11
Oxygen   106–110, 108t

P

Panope generosa   131, 132
Particles   6–7
Pectinaria californiensis   130
Pelletier, C.   24, 26
Periwinkles   144
pH   106–110, 108t
Pielou index   131
Pink salmon   151
Pisces IV   141
Pit lake   161–163, 162f, 163f
Plankton   105–106, 113. See also Zooplankton
   monitoring   154
Poling, G.W.   25, 26
Pollution Control Branch   19
   public hearing on DSTP at ICM   24–25, 59
Polychaete worm species   129–130
Protothaca staminea   154
PVC Niskin samplers   106
Pyrite   85, 88
Pyrrhotite   88

Q

Quadrat   203f
Quartz   85
Quartz Hill Mine (Alaska)   5
Quatsino Narrows   58, 66
   area maps   18f, 35f, 46f
   biological and chemical characterization   21
   physical characterization   21–23, 22f
   post-closure monitoring   164–165
   projected effects of tailing discharge   23–24
   sediment loss through   79
Quatsino Sound   18f, 21
   changes in metal levels 1971–1992   99–100,
      100t
   changes in physical and chemical properties
      93–94
   dissolved metals following mine closure
      100–102, 102t, 103
   field sampling   94
   mean oceanographic parameters before and after
      mining   95
   near-surface turbidity   95–99, 96f, 97t, 99t, 102
   oceanographic statistical analyses   94–95
   salinity   95, 102
   seawater sampling stations   91, 91f, 94, 94f
   Secchi depth   95–99, 96f, 98t, 99t
   temperature   95, 102

R

Reagents   42–43, 42*t*
Rescan Environmental Services Ltd.   25
Rhenium   40
Rockweed   143, 144
   increases in copper and zinc   154–156, 155*f*, 156
   monitoring   154
Rocky shore biodiversity   144, 204*f*
Royal BC Museum   119
Rupert Inlet   2–3, 17, 19. *See also* Morphological
      change (Rupert and Holberg Inlets)
   area maps   18*f*, 35*f*, 46*f*
   bathymetric surveys   47*t*
   biological and chemical characterization   21
   changes in metal levels 1971–1992   99–100,
      100*t*, 101*f*
   changes in physical and chemical properties
      93–94
   dissolved metals following mine closure
      100–102, 102*t*, 103
   elevated copper levels and biodiversity
      decline   154
   estimated deposition in   79
   field sampling   94
   fine-grained sediment   115–116
   geographical description   45
   mean oceanographic parameters before and after
      mining   95
   near-surface turbidity   95–99, 96*f*, 97*t*, 99*t*, 102
   oceanographic statistical analyses   94–95
   physical characterization   21–23, 22*f*
   post-closure monitoring   164–165
   pre-mine deposition patterns   59–60
   as primary recipient of tailings   165
   projected effects of tailing discharge   23–24
   salinity   95, 102
   seawater sampling stations   91, 91*f*, 94, 94*f*
   Secchi depth   95–99, 96*f*, 98*t*, 99*t*
   sediment troughs   116, 118*f*
   temperature   95, 102

S

SAG mills   38–39
Salinity   87, 95, 102
   post-closure   164
Salmon   90, 143
   data archives   148
   and government statistics   147
   hatchery   147, 151–152
   migrations   151
   monitoring   148, 151–152
Saxidomus giganteus   154
Sea cucumbers   130
Sea urchins   130
Seabed biodiversity   115–116, 164. *See also*
      Benthos, Intertidal biodiversity, Rocky shore
      biodiversity, Underwater biodiversity
   bivalve mollusks   129, 130, 132
   clams   131, 132
   crustacea   130–131, 132
   differing effects at different stations
      124–128
   diversity indices   131

Dungeness crab (*Cancer magister*)   130, 132
   echinoderms   130
   equilibrium concept vs. climax community
      128–129
   experimental colonization tests   127
   fish and invertebrates   131
   large infaunal species   131–132
   mollusks   132
   monitoring juveniles   132
   opportunist species   128, 129
   polychaete worm species   129–130
   and pore space   120
   post-closure   164–165
   sampling approach   116–120, 117*f*, 118*f*, 119*f*
   sediment grain size   120–123, 121*t*
   similarity analyses   131
   species evenness   128–130
   species richness   123–128, 125*t*–126*t*
   sustainable ecological succession   132
   thick tailings and reduction of species
      richness   124
Secchi disc   94, 95–99
Sediment
   accumulation and slime-worm   198*f*
   biodiversity   139
   fine-grained sediment in Rupert Inlet   115–116
   fish, crustacean, and quadrat on   203*f*
   grain size   120–123, 121*t*
   at Hankin Point   197*f*
   loss through Quatsino Narrows   79
   post-closure   164
   Rupert and Holberg Inlet troughs   116, 118*f*
   tailings and sedimentation rate   88
Semiautogenous grinding. *See* SAG mills
Settling plates   147, 157–159
   data   159, 159*t*
   sampling stations   157, 158*f*
Shannon index   131
Shrimp   141
SIGTREE software   131
Simon Fraser University   25
SIS RTM4002 digital thermometer   94
Slime-worm   198*f*
Slurry   1
   impoundments   1–2
Slurry pipelines   5–6
   up to 200 km   5, 11
Sockeye salmon   151
Sphalerite   88
Starfish   141, 143, 144
   at Hankin Point   198*f*, 199*f*
   on tailings outfall   202*f*
Statistical analyses   94–95
Stephensons' global rocky shore zonation
      scheme   144–145, 204*f*
*Submarine Disposal of Mill Tailings from On-Land
      Sources—An Overview and Bibliography*   3
Sulfate   89
Sulfate-reducing bacteria   163

T

Tailings   1
   acid-generating potential   86

bioassays   90, 147, 152–153, 153*t*
chemical and mineralogical compositions
   40–41, 41*t*
distribution of biogeochemically important
   species   89–90, 89*f*
fish and starfish on   202*f*
habitat change due to resuspension and
   upwelling   130
incorporation of organic detritus   88
mineralogical composition   85–86
particle size   86
permit limits   20, 20*t*, 21*t*
pH   87–88
post-closure   164
projected effects of discharge on receiving
   waters   23–24
redox potentials   88–90
salinity   87
and seawater column chemical quality   91, 91*f*,
   92*f*
and sediment grain size   120–123, 121*t*
and sedimentation rate   88
sulfide minerals   88
thick tailings and reduction of species
   richness   124
upwelling at Hankin Point   139, 159
zeta potential and seawater   87, 87*f*
Technical Environmental Advisory Committee
   26–27
Temperature
   correlations with zooplankton   110, 111*t*
   Rupert and Holberg Inlets and Quatsino
      Sound   95, 102
Terra Remote Sensing Inc.   51
Tharyx multifilis   129
Toquepala Mine (Peru)   2
Trace metal bioaccumulation   147, 150–151
   benthos tissue monitoring   153–157, 155*f*, 156*f*,
      157*f*
Tresus   132
Trout   151, 152
Tunicates   200*f*
Turbidity
   currents   6
   measurement system   94
   near-surface   95–99, 96*f*, 97*t*, 99*t*
   problems at scour hole   66, 69*f*, 70*f*
T. W. Beak Consultants   21

U

U.S. Borax   5
U.S. Bureau of Mines
   Alaska Field Operation Center   3–5
   publications on DSTP   3, 5
U.S. Environmental Protection Agency
   on DSTP   5
   on Quartz Hill Mine DSTP   5
U.S. Forest Service   5
UBC. *See* British Columbia, University of
Underwater biodiversity   137. *See also* Benthos,
      Rocky shore biodiversity, Intertidal biodiversity,
      Seabed biodiversity
   algal forests   139–140, 142

data archives   139
eelgrass   141
epifauna   142, 144
impact assessments   142
macrophytes   140
scuba surveys   137, 138*f*, 139
scuba surveys (1983, 1999)   142
scuba surveys by federal government
   scientists   140–141
scuba surveys by mine environmental staff and
   consultants   139–140
species on tailings   141
submersible surveys   137, 139, 141
Uniboom Seismic Profiler   59
Uppermost supra-littoral fringe   144
Utah Construction and Mining Company. *See* Utah
   Mining
Utah Mining   17–19, 59, 79

V

Varney Bay   140
Venables, William   19
Verrucaria   144
Victoria, University of   25, 139, 141, 148

W

Waldichuk-Buchanan report   26
Waste rock   36
Water contamination
   DSTP   11
   impoundments   2
Water recycling   12
Western Forest Products   151
Winter Harbour   58

Worms
   burrowing   21
   at Hankin Point   198*f*
   polychaete   129–130

Y

Yoldia scissurata   130
Yreka Mine   17

Z

Zinc
   in algae   154–156, 155*f*, 156
   correlations with zooplankton   110, 111*t*
   dissolved levels following mine closure
      100–102, 102*t*, 103
   levels in Dungeness crab   156, 157*f*
   monitoring   154
   in mussels   154, 155*t*, 156
   post-closure   164
   seawater levels 1971–1992   99–100, 100*t*
Zooplankton   105–106
   before and after mining   112–113, 113*t*
   correlations with chlorophyll A, temperature,
      dissolved copper, dissolved manganese,
      dissolved zinc, and total arsenic   110, 111*t*
   total   110–112, 110*t*, 111*t*, 112*t*
Zostera   140, 154

PLATE 9.1 Top: Two delicate, shell-less mollusks (nudibranchs) crawl together by an elongated tube-dwelling sea anemone on the rock at Hankin Point. The background consists of rocks with abundant algae and a patch of sediment. Bottom: A similar area at Hankin Point shows a film of sediment but with the same type of mollusk, a sea anemone, and the algal forest growing on the rocks. Both photographs were taken in July 1999.

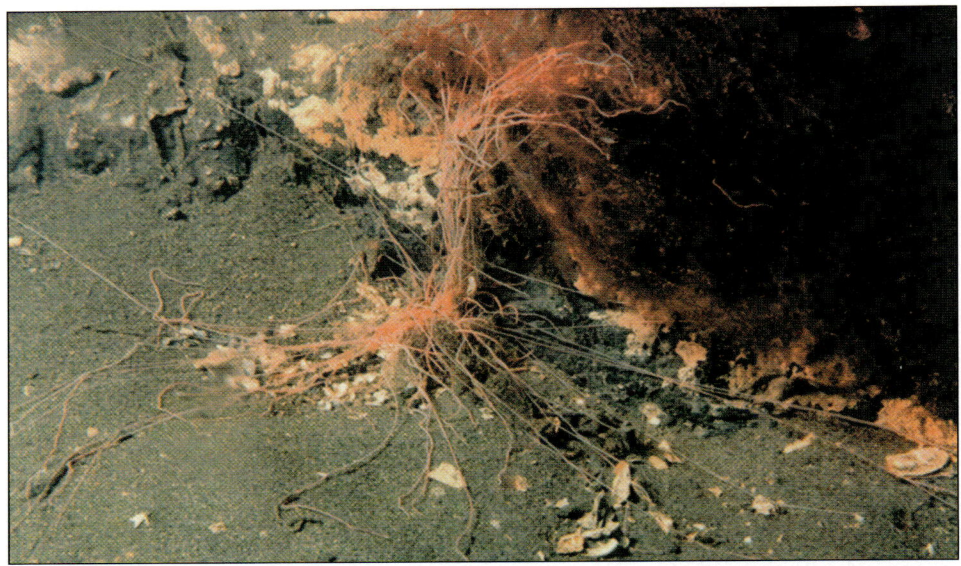

PLATE 9.2   Top: This surface roughness feature has accumulated sediment where a massive slime-worm is growing. Dense algae grow on the rocks that remain exposed. Bottom: On this patch of sediment between algae-supporting rocks is one of the many fish recorded, along with a starfish that is apparently feeding on a prey organism. Both photographs were taken in July 1999 at Hankin Point.

PLATE 9.3 Top: This patch of sediment at Hankin Point is supporting two large burrowing sea anemones and a starfish. Bottom: A Dungeness crab is burrowed into the patch of sediment, with algae growing on the nearby exposed rock. Both photographs were taken in July 1999.

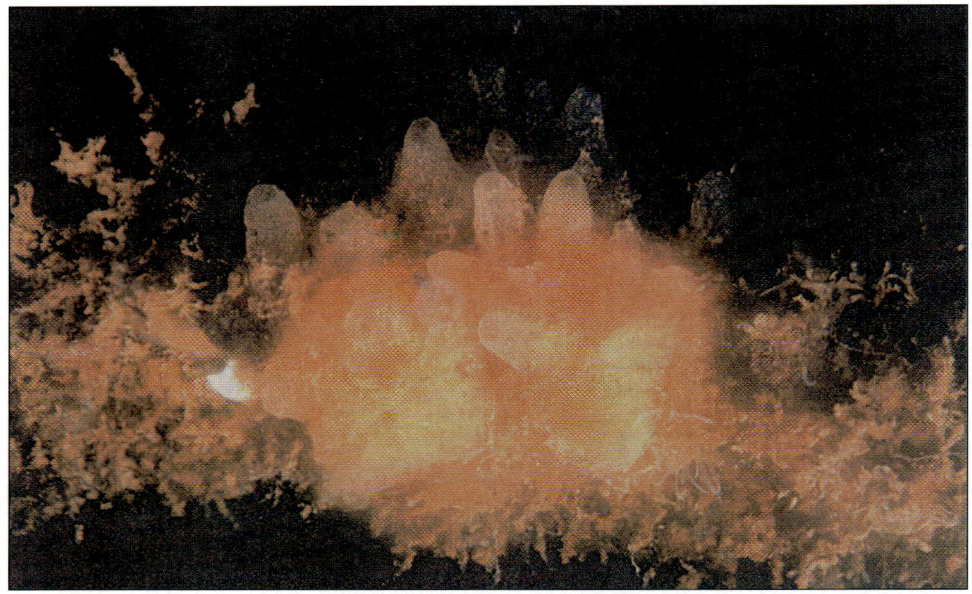

PLATE 9.4 Top: A supportive cable at the mine dock is crowded with growth, including a patch of tunicates, a great deal of algae, and encrusting organisms. Bottom: In the near-shore mud around the dock, substantial numbers of burrowing sea anemones are living. The veneer of brown organic ooze common on muds throughout the fjord system is visible. Both photographs were taken in July 1999.

PLATE 9.5 Top: Thick and diverse encrusting organisms, including one large tubeworm that is well extended and feeding, grow on the tailings outfall. **Bottom:** A dense colony of sea anemones is growing on the outfall. Both photographs were taken in July 1999.

PLATE 9.6 Top: A fish grazes on the dense encrusting growth on the tailings outfall. Bottom: On the tailings bed by the outfall site, a starfish moves past the imprint of a smaller starfish in the brown organic ooze widely spread throughout the fjord system. Both photographs were taken in July 1999.

PLATE 9.7 Top: The view from the submersible port over natural (brown) sediments showing the 1-m^2 quadrat and the measuring rope. A crustacean is present in the foreground by the rope, with several burrow holes nearby. Bottom: A similar view over the tailings (gray). A small fish is present at left center, and several burrow holes are visible.

PLATE 9.8 Top: A large boulder near the mine site shows the typical regional shoreline zonation pattern, with its expression on the surrounding stony beach. Bottom: The shoreline front between normal rock shore zonation for a stony/boulder beach (left) and the first-year colonization of low-tide green and brown filamentous algae (right).